SIXTH EDITION

Atlas of Skeletal Muscles

Robert J. Stone
Suffolk County Community College

Judith A. Stone
Suffolk County Community College

**McGraw-Hill
Higher Education**

Boston   Burr Ridge, IL   Dubuque, IA   New York   San Francisco   St. Louis
Bangkok   Bogotá   Caracas   Kuala Lumpur   Lisbon   London   Madrid   Mexico City
Milan   Montreal   New Delhi   Santiago   Seoul   Singapore   Sydney   Taipei   Toronto

The McGraw·Hill Companies

McGraw-Hill
Higher Education

ATLAS OF SKELETAL MUSCLES, SIXTH EDITION

Published by McGraw-Hill, a business unit of The McGraw-Hill Companies, Inc., 1221 Avenue of the Americas, New York, NY 10020. Copyright © 2009 by The McGraw-Hill Companies, Inc. All rights reserved. Previous editions © 2006, 2003, 2000, 1997 and 1990. No part of this publication may be reproduced or distributed in any form or by any means, or stored in a database or retrieval system, without the prior written consent of The McGraw-Hill Companies, Inc., including, but not limited to, in any network or other electronic storage or transmission, or broadcast for distance learning.

Some ancillaries, including electronic and print components, may not be available to customers outside the United States.

This book is printed on recycled, acid-free paper containing 10% postconsumer waste.

1 2 3 4 5 6 7 8 9 0 QPD/QPD 0 9 8

ISBN 978–0–07–128359–5
MHID 0–07–128359–5

www.mhhe.com

Dedication

To Karen, Andrew, and Laura, helping us more and more as the years go by.

Contents

CHAPTER SEVEN

Muscles of the Forearm and Hand 129

CHAPTER EIGHT

Muscles of the Hip and Thigh 163

CHAPTER NINE

Muscles of the Leg and Foot 193

Alphabetical Listing of Muscles 221

Index 223

Preface

This book is a study guide and reference for the anatomy and actions of human skeletal muscles. It is designed for use by students of anatomy, physical education, and health-related fields. It also serves as a compact reference for the practicing professional.

The first chapter presents photographic illustrations of the major features of the skeleton. These photos have been selectively enhanced and combined to emphasize important features. They are thus a hybrid between drawings and unretouched photographs. The philosophy embraced is that the teacher function as a lens to focus attention on selected facts and observations. A master numbering system is used so that each structure is labeled with the same number in all drawings.

The second chapter describes through illustration and description the various movements of the body.

In chapters 3 through 9, the origin, insertion, action, and innervation of the skeletal muscles are described, and each muscle is presented on a separate page with a line drawing.

The spinal cord levels of the nerve fibers that innervate each muscle are included in parentheses after the name of each nerve.

Labeled drawings of major muscle groups are presented throughout chapters 3 through 9. Notes and relationships among muscles have been included on many pages.

The drawings include the following important features:

1. Bones and cartilage containing muscle attachments are shaded.
2. Adjacent structures are shown.
3. Muscle fibers are drawn by direction.
4. Muscle fibers are shown on the undersurface of bone and cartilage as dashed lines.
5. Tendons and aponeuroses are shown.
6. Labeled muscle groups are included.

These features aid in visual orientation and understanding of the action of the muscles. We have noticed that many students find it useful to color the illustrations.

Notes have been included on many pages to show how muscles are used. Relationships among many of the muscles have also been indicated where appropriate.

Some users of previous editions have advised that some of the smaller muscles should be enlarged and shown with less skeletal background. We have purposely standardized the skeletal views to allow an appreciation of the relative sizes and positions of the muscles. Skeletal muscles, at the gross level, are relatively simple anatomical structures, so very little additional information would be included by enlargement, and many comparative relationships would be lost.

Our primary objective is to describe the muscles moving the skeleton, therefore we have not included the muscles of the peritoneum, eye, tympanic cavity, tongue, larynx, pharynx, or palate.

We extend our appreciation to Mr. George Boykin, for many years the jolly proprietor of the gross anatomy laboratories at the State University of New York at Stony Brook, for his help and encouragement. We also thank Mr. Vincent Verdisco for his technical advice and the many students who have offered valuable suggestions over the years.

We would also like to thank the many reviewers who have made helpful suggestions for improving past editions of this atlas, as well as Leann Blem, *University of Montana;* Christine A. Byrd-Jacobs, *Western Michigan University;* Pamela B. Fouche, *Walters State Community College;* Candice Francis, *Palomar College;* Michael Hendrix, *Missouri State University;* Steven L. Keffer, *James Madison University;* Dennis Landin, *Louisiana State University;* Malinda McMurry, *Morehead State University;* Virginia L. Naples, *Northern Illinois University;* Donna Newhouse, *Lakehead University;* Russell Nordeen, *University of Arkansas at Monticello;* David Pearson, *Ball State University;* Mark L. Wagner, *CollegeAmerica* for their input on the sixth edition.

Robert J. Stone

Judith A. Stone

CHAPTER ONE

The Skeleton

SKULL—ANTERIOR VIEW

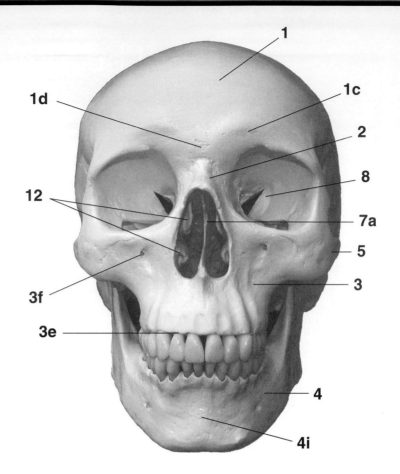

1. Frontal bone
1c. Superciliary arch
1d. Glabella
2. Nasal bone
3e-h. Maxilla
3e. Alveolar border
3f. Infraorbital foramen
3g. Incisive foramen
3h. Palatine process
4. Mandible
4i. Symphysis
5. Zygomatic bone
7a. Perpendicular plate of ethmoid bone
8. Sphenoid
8b. Lateral pterygoid plate

8d. Medial pterygoid plate
8e. Foramen ovale
9b-f. Temporal bone
9b. Mastoid process (temporal bone)
9c. Foramen lacerum
9e. Zygomatic process (temporal bone)
9f. Carotid canal
11a. Superior nuchal line (occipital bone)
11b. Inferior nuchal line (occipital bone)
11c. External occipital protuberance (occipital bone)
11d. Occipital condyle (occipital bone)
11e. External occipital crest (occipital bone)
11f. Jugular foramen
11g. Foramen magnum
12. Turbinates
13. Palatine bone
14. Vomer

SKULL—INFERIOR (BASAL) VIEW

SKULL—LATERAL VIEW

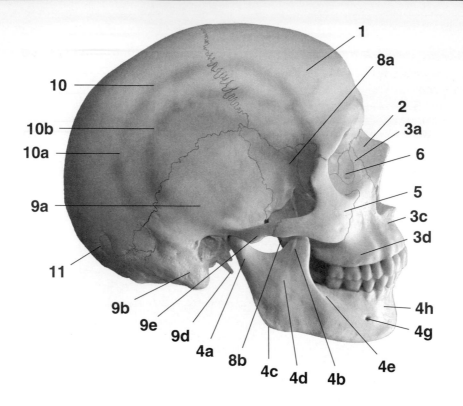

1. Frontal bone	**9a-e.** Temporal bone
2. Nasal bone	**9a.** Temporal fossa (squamous part of temporal bone)
3a-d. Maxilla	**9b.** Mastoid process
3a. Frontal process	**9d.** Styloid process
3c. Incisive process	**9e.** Zygomatic process
3d. Canine fossa	**10.** Parietal bone
4a-h. Mandible	**10a.** Superior temporal line
4a. Neck of condyle	**10b.** Inferior temporal line
4b. Coronoid process	**11.** Occipital bone
4c. Angle	
4d. Ramus	Note: The zygomatic arch is formed by the zygomatic process of
4e. Oblique line	the temporal bone meeting the zygomatic bone.
4g. Mental foramen	
4h. Incisive fossa	
5. Zygomatic bone	
6. Lacrimal bone	
8a. Greater wing of sphenoid bone	
8b. Lateral pterygoid plate of sphenoid bone	

VERTEBRAL COLUMN—LATERAL VIEW

C1-7. Cervical vertebrae
C1. Atlas
C2. Axis
C7. Seventh cervical vertebra
T1-12. Thoracic vertebrae
T1. First thoracic vertebra
T12. Twelfth thoracic vertebra
L1-5. Lumbar vertebrae
L1. First lumbar vertebra
L5. Fifth lumbar vertebra
S. Sacrum
Co. Coccyx

ATLAS AND AXIS

Atlas

1. Anterior tubercle
2. Anterior arch
3. Superior articular facet
4. Posterior arch
5. Posterior tubercle
6. Transverse foramen

Axis

7. Dens (odontoid process)
8. Spinous process
9. Transverse process

THORACIC VERTEBRAE—LATERAL VIEW

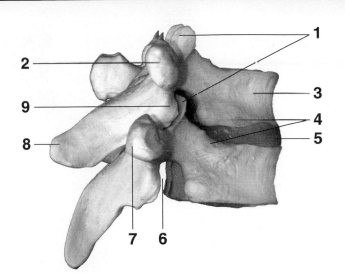

1. Superior articular process
2. Transverse process
3. Body
4. Demifacets (for ribs)
5. Disk
6. Inferior vertebral notch
7. Facet (for rib tubercle)
8. Spinous process
9. Inferior articular process

LUMBAR VERTEBRA—SUPERIOR VIEW

1. Spinous process
2. Mammillary process
3. Transverse process
4. Lamina
5. Vertebral foramen
6. Pedicle
7. Body (centrum)

SACRUM—POSTERIOR VIEW

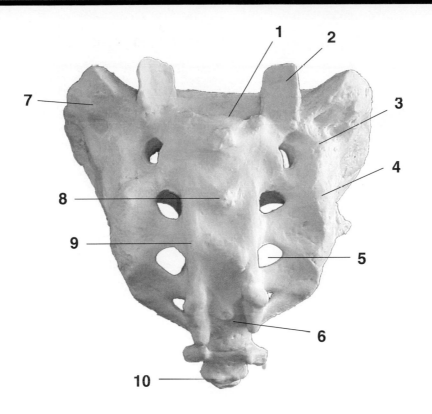

1. Sacral canal
2. Superior articular facet
3. Sacral tuberosity
4. Lateral crest
5. Posterior sacral foramen
6. Sacral hiatus (opening of sacral canal)
7. Auricular surface
8. Median crest
9. Intermediate crest
10. Coccyx

SACRUM—PELVIC VIEW

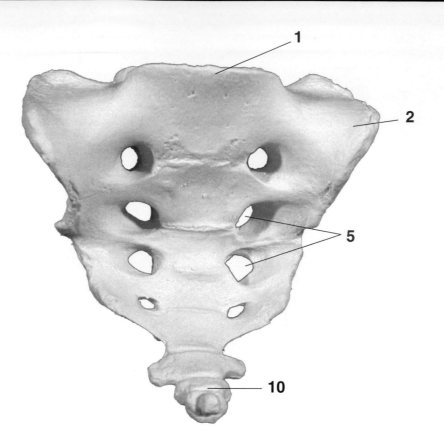

1. Promontory
2. Ala
5. Sacral foramina
10. Coccyx

STERNUM AND CLAVICLE WITH SCAPULA

Clavicle
P = proximal end (sternal)
D = distal end (acromial or lateral)
Scapula (S)
A = acromion
Sternum
M = manubrium
SA = sternal angle
B = body
X = xiphoid process

The proximal (sternal) end of the clavicle forms the sternoclavicular joint with the manubrium of the sternum.

The distal (scapular) end forms the acromioclavicular joint with the acromion of the scapula. This is the only bony articulation of the upper limb with the torso.

The second rib cartilage articulates at the sternal angle between the manubrium and the body of the sternum.

The cartilages of ribs 7-10 are fused to form the costal arch.

HYOID BONE—SUPERIOR VIEW

18a-c. Hyoid bone
 18a. Greater horn
 18b. Body
 18c. Lesser horn

The greater horn of the hyoid bone attaches to the styloid processes of the temporal bones by the styloid ligaments. It attaches to the thyroid cartilage by the thyrohyoid ligament and supports the upper respiratory tract.

RIB ARTICULATIONS

The head (H) of a rib articulates with the demifacets of two adjacent vertebrae. The tubercle (T) joins the facet of the transverse process of the upper vertebra. The anterior end (A) meets the costal cartilage, which then joins the sternum.

SCAPULA AND HUMERUS—ANTERIOR VIEW

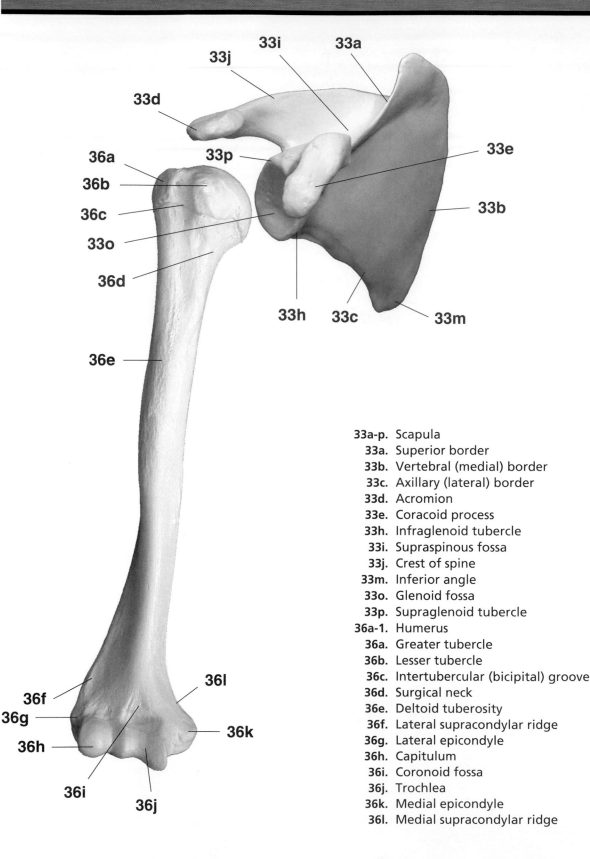

33a-p. Scapula
  33a. Superior border
  33b. Vertebral (medial) border
  33c. Axillary (lateral) border
  33d. Acromion
  33e. Coracoid process
  33h. Infraglenoid tubercle
  33i. Supraspinous fossa
  33j. Crest of spine
  33m. Inferior angle
  33o. Glenoid fossa
  33p. Supraglenoid tubercle
36a-1. Humerus
  36a. Greater tubercle
  36b. Lesser tubercle
  36c. Intertubercular (bicipital) groove
  36d. Surgical neck
  36e. Deltoid tuberosity
  36f. Lateral supracondylar ridge
  36g. Lateral epicondyle
  36h. Capitulum
  36i. Coronoid fossa
  36j. Trochlea
  36k. Medial epicondyle
  36l. Medial supracondylar ridge

SCAPULA AND HUMERUS—POSTERIOR VIEW

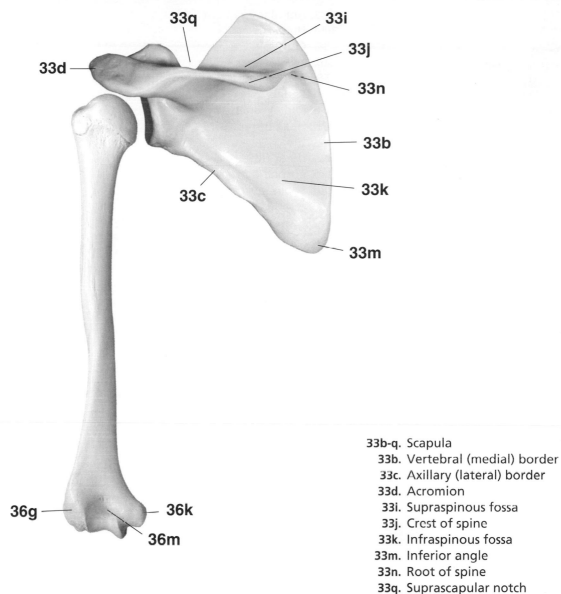

33b-q. Scapula
  33b. Vertebral (medial) border
  33c. Axillary (lateral) border
  33d. Acromion
  33i. Supraspinous fossa
  33j. Crest of spine
  33k. Infraspinous fossa
  33m. Inferior angle
  33n. Root of spine
  33q. Suprascapular notch
36g-m. Humerus
  36g. Lateral epicondyle
  36k. Medial epicondyle
  36m. Olecranon fossa

ELBOW—ANTERIOR VIEW

35. Annular ligament
36i-l. Humerus
  36i. Coronoid fossa
  36k. Medial epicondyle
  36l. Medial supracondylar ridge
37a, b. Radius
  37a. Radial tuberosity
  37b. Head of radius
38a-d. Ulna
  38a. Ulnar tuberosity
  38b. Coronoid process
  38d. Supinator crest

FOREARM—POSTERIOR VIEW

34. Radial (lateral) collateral ligament
35. Annular ligament
36k-m. Humerus
36k. Medial epicondyle
36m. Olecranon fossa
37. Radius
37d. Styloid process of radius
38c-e. Ulna
38c. Olecranon
38e. Styloid process
39. Interosseous membrane

HAND—PALMAR VIEW

40. Scaphoid
40a. Tubercle of scaphoid
41. Lunate
42. Pisiform
43a. Tubercle of trapezium
44. Trapezoid
45. Capitate
46a. Hook of hamate
47. Triquetrum
48. Metacarpals

49a. Proximal (first) phalanges
49b. Middle (second) phalanges
49c. Distal (third) phalanges
50. Flexor retinaculum
51. Palmar aponeurosis

HAND—DORSAL VIEW

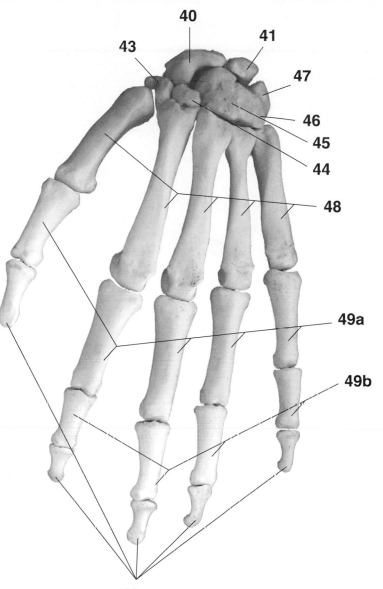

40. Scaphoid
41. Lunate
43. Trapezium
44. Trapezoid
45. Capitate
46. Hamate
47. Triquetrum
48. Metacarpals

49a. Proximal (first) phalanges
49b. Middle (second) phalanges
49c. Distal (third) phalanges

PELVIS—ANTERIOR VIEW

52b-i. Ilium
52b. Iliac crest
52c. Iliac tubercle
52e. Anterior superior iliac spine
52f. Anterior inferior iliac spine
52g. Iliac fossa
52h. Arcuate line
52i. Iliopectineal eminence
53. Ischium
54a-e. Pubis
54a. Superior ramus
54b. Inferior ramus
54c. Pubic crest
54d. Pubic symphysis
54e. Pubic tubercle
57. Obturator foramen
63. Acetabulum
68a-e. Femur
68a. Greater trochanter
68b. Trochanteric fossa
68c. Lesser trochanter
68e. Intertrochanteric line

PELVIS—THREE QUARTER POSTERIOR VIEW

52m
52j
52n
55
56
53a
53c
53d
53
53b
68c
68

52k
52l
68a
68d
68g
68h

52j-n. Ilium
  52j. Posterior gluteal line
  52k. Middle gluteal line
  52l. Inferior gluteal line
  52m. Posterior superior iliac spine
  52n. Posterior inferior iliac spine
53a-d. Ischium
  53a. Sacrospinous ligament
  53b. Ischial tuberosity
  53c. Ischial spine
  53d. Lesser sciatic notch (lesser sciatic foramen)*
  55. Greater sciatic notch (greater sciatic foramen)*
  56. Sacrotuberous ligament
68a-h. Femur
  68a. Greater trochanter
  68c. Lesser trochanter
  68d. Intertrochanteric crest
  68g. Gluteal tuberosity
  68h. Linea aspera

*Note: These bony notches are converted to foramina by the two ligaments.

FEMUR—POSTERIOR VIEW

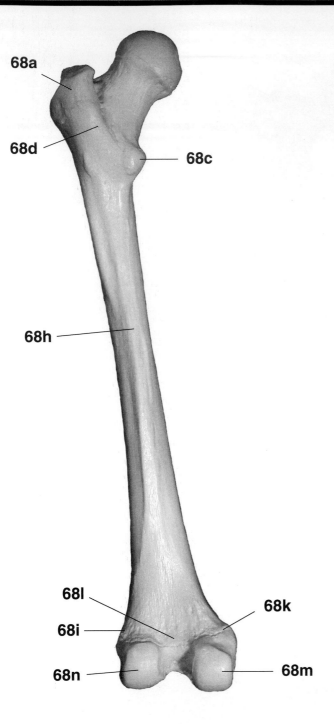

68a-n. Femur
68a. Greater trochanter
68c. Lesser trochanter
68d. Intertrochanteric crest
68h. Linea aspera
68i. Lateral supracondylar ridge
68k. Adductor tubercle
68l. Popliteal surface
68m. Medial condyle
68n. Lateral condyle

KNEE—ANTERIOR VIEW

68i-k. Femur
  68i. Lateral supracondylar ridge
  68j. Medial supracondylar ridge
  68k. Adductor tubercle
  70. Patella
  71. Patellar ligament
74a-c. Tibia
  74a. Tibial tuberosity
  74b. Lateral condyle
  74c. Medial condyle
  75a. Head of fibula

Note: The quadriceps tendon continues through the patella and becomes the patellar ligament.

ANKLE AND FOOT—ANTEROLATERAL VIEW

74. Tibia
74d. Medial malleolus of tibia
75. Fibula
76. Interosseous membrane
77. Lateral talocalcaneal ligament
78. Inferior extensor retinaculum
83. Talus
84. Calcaneus
85. Navicular
86. Medial cuneiform
87. Intermediate cuneiform
88. Lateral cuneiform
89. Cuboid
90a-e. Metatarsal bones
90a. First metatarsal
90b. Second metatarsal
90c. Third metatarsal
90d. Fourth metatarsal
90e. Fifth metatarsal
91a. Proximal phalanges
91b. Middle phalanges
91c. Distal phalanges

FOOT—PLANTAR VIEW

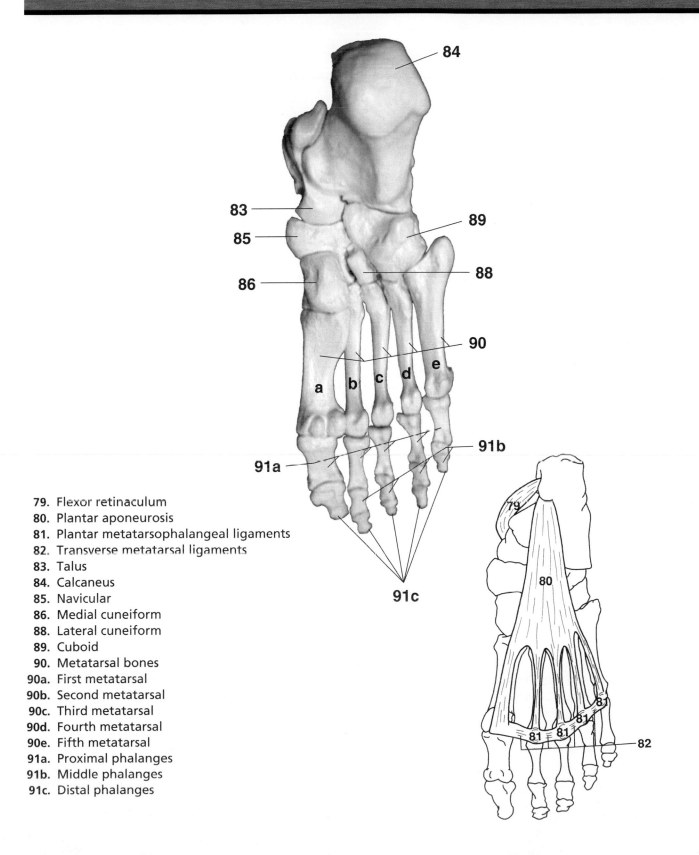

79. Flexor retinaculum
80. Plantar aponeurosis
81. Plantar metatarsophalangeal ligaments
82. Transverse metatarsal ligaments
83. Talus
84. Calcaneus
85. Navicular
86. Medial cuneiform
88. Lateral cuneiform
89. Cuboid
90. Metatarsal bones
90a. First metatarsal
90b. Second metatarsal
90c. Third metatarsal
90d. Fourth metatarsal
90e. Fifth metatarsal
91a. Proximal phalanges
91b. Middle phalanges
91c. Distal phalanges

CHAPTER TWO

Movements
of the Body

Anatomical position—A subject in the anatomical position is standing erect with the head, eyes, and toes facing forward and the arms hanging straight at the sides with the palms of the hands facing forward.

FIGURE 2.2
Coronal (frontal) planes—Pass vertically through the body from side to side. They divide the body from front to back.

FIGURE 2.1
Median or midsagittal plane—Passes vertically through the body from anterior (front) to posterior (back). It divides the body into right and left sides. Other sagittal planes are parallel to this plane.

FIGURE 2.3
Transverse planes (cross sections)—Pass horizontally through the body parallel to the ground.

FIGURE 2.4
Flexion-extension—Starting from the anatomical position, movement is anterior or posterior in a sagittal plane. In the hinge joints *flexion* results in a decrease in the angle made by the bones in the joint, and *extension* brings the bones toward a 180° angle. Hyperextension is permitted in the shoulder, wrist, and vertebrae where movement continues posterior to anatomical position. In the knee, hip, and elbow, hyperextension is prevented by bone structure or ligaments.

FIGURE 2.5
Lateral flexion—The torso (or head) bends laterally in the coronal plane.

FIGURE 2.7
Medial rotation—The anterior of the arm (or thigh) is moved toward the median plane.
Lateral rotation—The anterior of the arm (or thigh) is moved away from the median plane.

FIGURE 2.6
Abduction—The right leg moves laterally in the coronal plane.
Adduction—The leg is returned medially in the coronal plane.

Note: If the foot is fixed, this motion results in tilting the pelvis upward on the opposite side.

MOVEMENTS OF THE SCAPULA

FIGURE 2.8
Elevation—The right
scapula of this figure is
drawn superiorly against
the resistance of the rock.

FIGURE 2.9
Depression—The right
scapula of this figure is
pushing the arm inferiorly.

FIGURE 2.10
Protraction—The scapula pushes the arm forward
in a sagittal plane.

FIGURE 2.11
Retraction—The scapula is pulled back from protraction
in a sagittal plane. The scapula slides around the ribs
toward the median plane, so it becomes adduction.

FIGURE 2.12
Rotation—For abduction of the arm to continue above
the height of the shoulder, the scapula must rotate on
its axis so that the glenoid fossa turns upward.

MOVEMENTS OF THE HAND AND FOREARM

FIGURE 2.14
Abduction—The fingers are moved away from the mid-line of the hand.

FIGURE 2.13
Pronation—The forearm is rotated away from the anatomical position so that the palm turns medially, then posteriorly. If the forearm is flexed at the elbow, then the palm turns inferiorly.
Supination—The forearm is rotated so that the palm turns anteriorly (or superiorly if the forearm is flexed). Also see figure 2.19.

FIGURE 2.16
Adduction—The fingers are moved toward the midline of the hand.

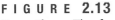

FIGURE 2.15
Radial deviation (abduction)—The hand, at the wrist, is moved laterally toward the radius. In anatomical position this moves the hand away from the body in the coronal plane.
Ulnar deviation (adduction)—The hand, at the wrist, is moved medially toward the ulna.

FIGURE 2.17
Opposition—The saddle joint between the trapezium and first metacarpal allows the thumb to turn so its anterior surface can touch the anterior surfaces of the four fingers when they are partially flexed.

MOVEMENTS OF THE FOOT

FIGURE 2.18
Dorsiflexion—Elevating the foot, decreasing the angle between the foot and the leg.
Plantar flexion—Depressing the foot, increasing the angle between the foot and the leg.

FIGURE 2.19
Inversion-eversion—During *inversion* the sole of the foot turns medially, and during *eversion* it returns or turns slightly laterally. Most of this movement is permitted by the gliding of the intertarsal and tarsometatarsal joints but because the articulating surface of the talus is narrower posteriorly, it permits some lateral movement when the ankle is plantar flexed.

CHAPTER THREE

Muscles of the Face and Head

EPICRANIUS

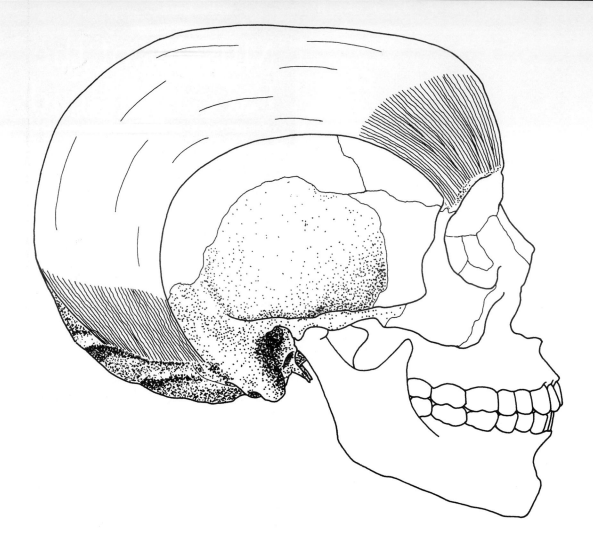

Skull—lateral view

Occipital belly *(occipitalis)*

■ **Origin**
Lateral two-thirds of superior nuchal line of occipital bone, mastoid process of temporal bone

■ **Insertion**
Galea aponeurotica (an intermediate tendon leading to frontal belly)

■ **Action**
Draws back scalp, aids frontal belly to wrinkle forehead and raise eyebrows

■ **Nerve**
Posterior auricular branch of facial nerve

Frontal belly *(frontalis)*

■ **Origin**
Galea aponeurotica

■ **Insertion**
Fascia of facial muscles and skin above nose and eyes

■ **Action**
Draws back scalp, wrinkles forehead, raises eyebrows

■ **Nerve**
Temporal branches of facial nerve

TEMPOROPARIETALIS

Skull—lateral view

■ **Origin**
Fascia over ear

■ **Insertion**
Lateral border of galea aponeurotica

■ **Action**
Raises ears, tightens scalp

■ **Nerve**
Temporal branch of facial nerve

AURICULARIS ANTERIOR, SUPERIOR, POSTERIOR

Skull—lateral view

Auricularis anterior

■ Origin
Fascia in temporal region

■ Insertion
Anterior to helix of ear

■ Action
Draws ear forward in some individuals, moves scalp*

■ Nerve
Temporal branch of facial nerve

Auricularis superior

■ Origin
Fascia in temporal region

■ Insertion
Superior part of ear

■ Action
Draws ear upward in some individuals, moves scalp*

■ Nerve
Temporal branch of facial nerve

Auricularis posterior

■ Origin
Mastoid area of temporal bone

■ Insertion
Posterior part of ear

■ Action
Draws ear upward in some individuals*

■ Nerve
Posterior auricular branch of facial nerve

*This muscle is nonfunctional in most people.

ORBICULARIS OCULI

ORBITAL AND PALPEBRAL PARTS

LACRIMAL PART

Skull—three-quarter anterior view

Orbital part

■ **Origin**
Frontal bone, maxilla (medial margin of orbit)

■ **Insertion**
Continues around orbit and returns to origin

■ **Action**
Strong closure of eyelids

■ **Nerve**
Temporal and zygomatic branches of facial nerve

Palpebral part *(in eyelids)*

■ **Origin**
Medial palpebral ligament

■ **Insertion**
Lateral palpebral ligament into zygomatic bone

■ **Action**
Gentle closure of eyelids

■ **Nerve**
Temporal and zygomatic branches of facial nerve

Lacrimal part *(behind medial palpebral ligament and lacrimal sac)*

■ **Origin**
Lacrimal bone

■ **Insertion**
Lateral palpebral raphe

■ **Action**
Draws lacrimal canals onto surface of eye

■ **Nerve**
Temporal and zygomatic branches of facial nerve

LEVATOR PALPEBRAE SUPERIORIS

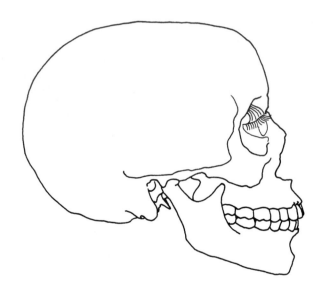

Skull—three-quarter anterior view

■ Origin
Inferior surface of lesser wing of sphenoid

■ Insertion
Skin of upper eyelid

■ Action
Raises upper eyelid

■ Nerve
Oculomotor nerve

Skull—lateral view

CORRUGATOR SUPERCILII

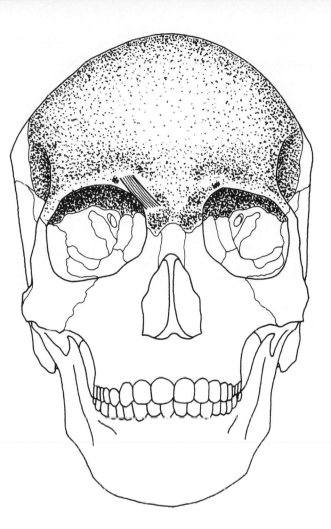

Skull—anterior view

■ **Origin**
Medial end of superciliary arch

■ **Insertion**
Deep surface of skin under medial portion of eyebrows

■ **Action**
Draws eyebrows downward and medially

■ **Nerve**
Temporal branch of facial nerve

PROCERUS

Skull—anterior view

■ Origin
Fascia over nasal bone and lateral nasal cartilage

■ Insertion
Skin between eyebrows

■ Action
Draws down medial part of eyebrows, wrinkles nose

■ Nerve
Buccal branches of facial nerve

NASALIS

Skull—anterior view

Transverse part

■ **Origin**
Middle of maxilla

■ **Insertion**
Muscle of opposite side over bridge of nose

Alar part

■ **Origin**
Greater alar cartilage, skin on nose

■ **Insertion**
Skin at point of nose

■ **Action**
Both parts maintain opening of external nares during forceful inspiration

■ **Nerve**
Buccal branches of facial nerve

DEPRESSOR SEPTI

Skull—anterior view

■ **Origin**
Incisive fossa of maxilla

■ **Insertion**
Nasal septum and ala

■ **Action**
Constricts nares

■ **Nerve**
Buccal branches of facial nerve

ORBICULARIS ORIS

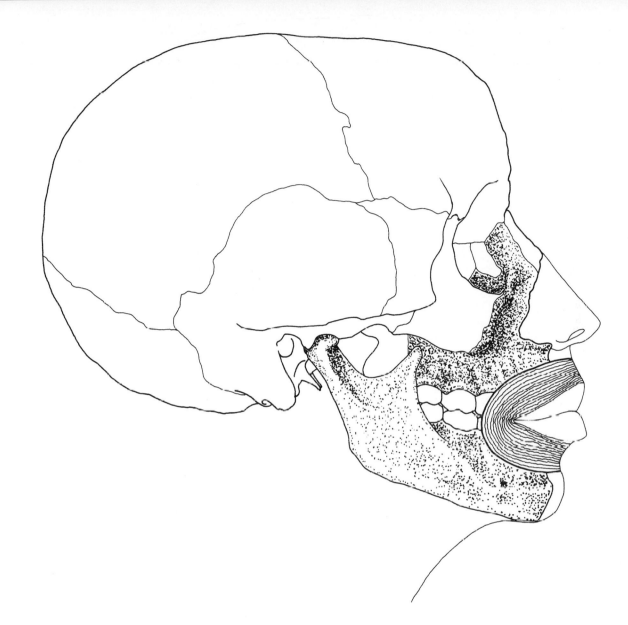

Skull—lateral view

■ Origin
Lateral band—alveolar border of maxilla

Medial band—septum of nose

Inferior portion—lateral to midline of mandible

■ Insertion
Becomes continuous with other muscles at angle of mouth

■ Action
Closure and protrusion of lips

■ Nerve
Buccal and mandibular branches of facial nerve

LEVATOR LABII SUPERIORIS

Skull—lateral view

Angular head (*aleque nasi*)

■ Origin
Frontal process of maxilla and zygomatic bone

■ Insertion
Greater alar cartilage and skin of nose, upper lip

■ Action
Elevates upper lip, dilates nares, forms nasolabial furrow

■ Nerve
Buccal branches of facial nerve

Infraorbital head

■ Origin
Infraorbital margin of maxilla

■ Insertion
Skin of lateral half of upper lip

■ Action
Elevates upper lip, forms nasolabial furrow

■ Nerve
Buccal branches of facial nerve

Note: The angular head is frequently referred to as a separate muscle, levator labii superioris aleque nasi.

LEVATOR ANGULI ORIS

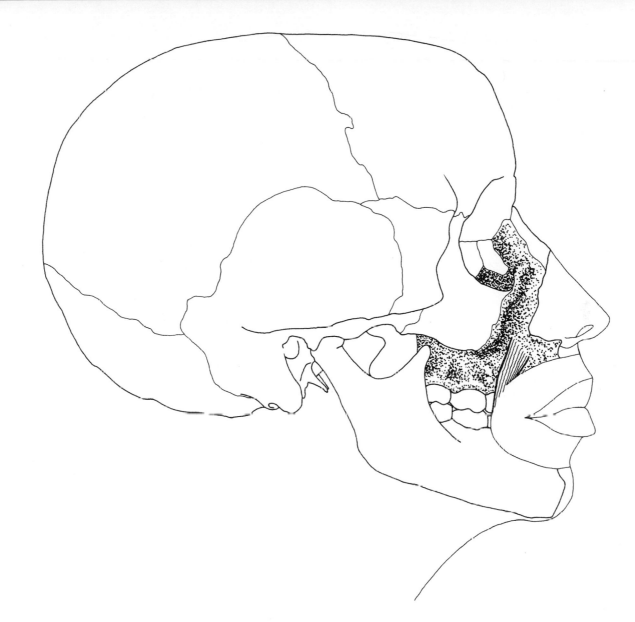

Skull—lateral view

■ **Origin**
Canine fossa of maxilla

■ **Insertion**
Angle of mouth

■ **Action**
Elevates corner (angle) of mouth

■ **Nerve**
Buccal branches of facial nerve

ZYGOMATICUS MAJOR

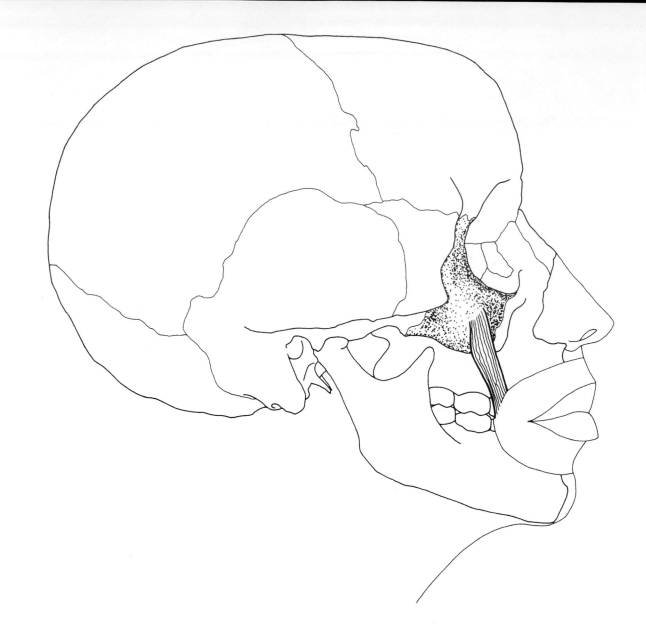

Skull—lateral view

- **Origin**
Zygomatic bone

- **Insertion**
Angle of mouth

- **Action**
Draws angle of mouth upward and backward (laughing)

- **Nerve**
Buccal branches of facial nerve

ZYGOMATICUS MINOR

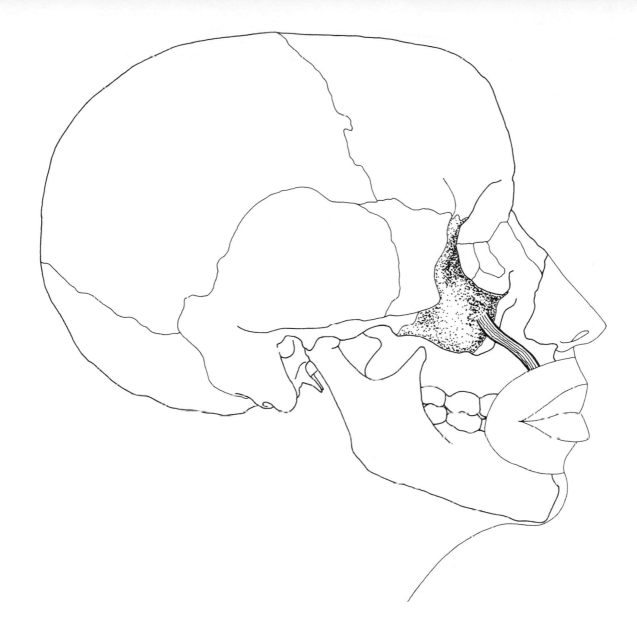

Skull—lateral view

■ **Origin**
Zygomatic bone

■ **Insertion**
Upper lip lateral to levator labii superioris

■ **Action**
Forms nasolabial furrow

■ **Nerve**
Buccal branches of facial nerve

RISORIUS

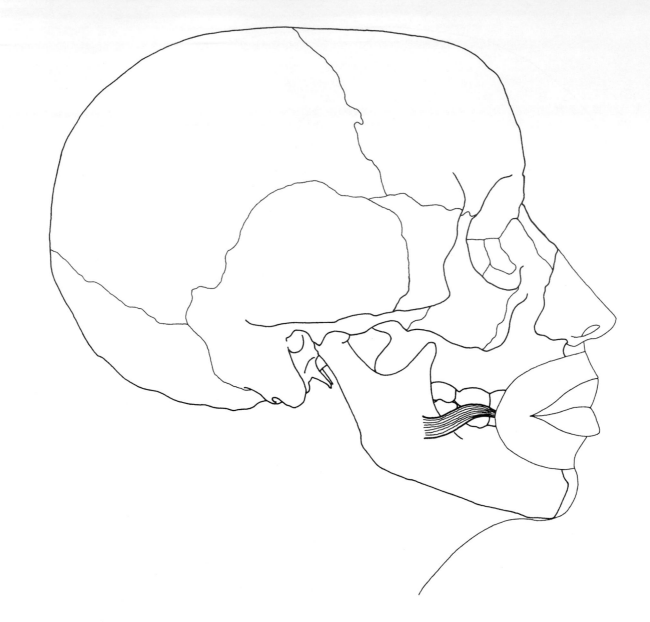

Skull—lateral view

■ Origin
Fascia over masseter

■ Insertion
Skin at angle of mouth

■ Action
Retracts angle of mouth, as in grinning

■ Nerve
Buccal branches of facial nerve

DEPRESSOR LABII INFERIORIS

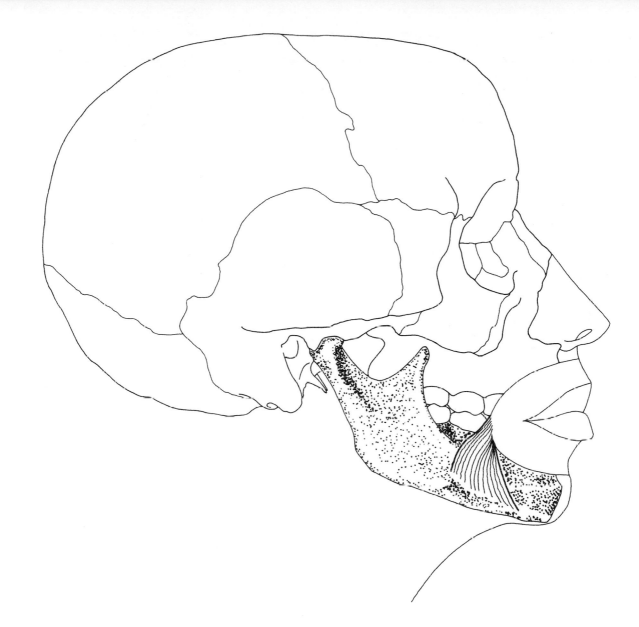

Skull—lateral view

■ **Origin**
Mandible, between symphysis and mental foramen

■ **Insertion**
Skin of lower lip

■ **Action**
Draws lower lip downward and laterally

■ **Nerve**
Mandibular branch of facial nerve

DEPRESSOR ANGULI ORIS

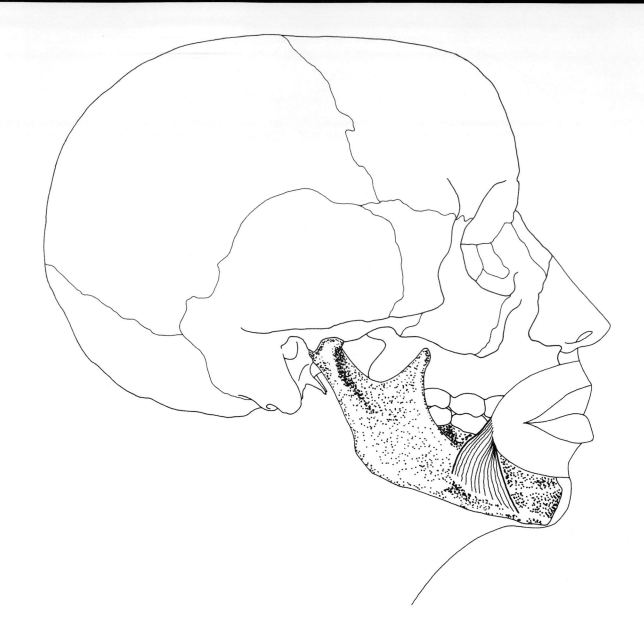

Skull—lateral view

■ Origin
Oblique line of the mandible

■ Insertion
Angle of the mouth

■ Action
Depresses angle of mouth, as in frowning

■ Nerve
Mandibular branch of facial nerve

MENTALIS

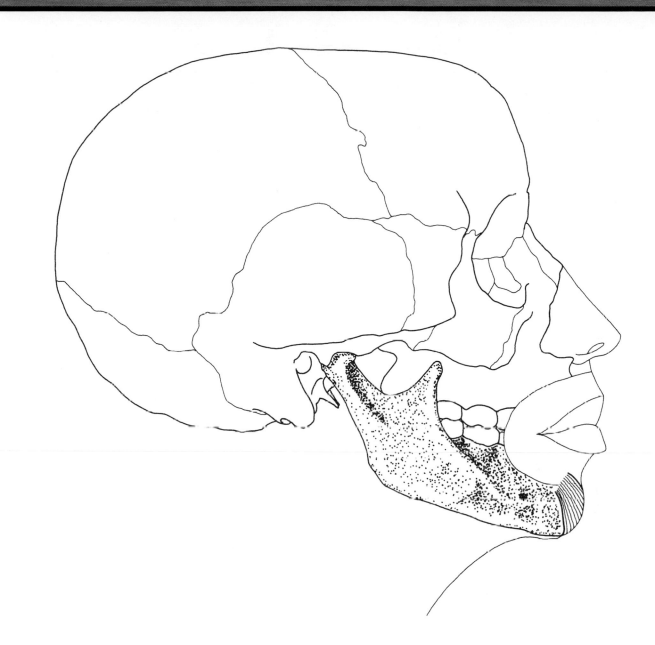

Skull—lateral view

■ **Origin**
Incisive fossa of mandible

■ **Insertion**
Skin of chin

■ **Action**
Raises and protrudes lower lip, wrinkles skin of chin

■ **Nerve**
Mandibular branch of facial nerve

BUCCINATOR

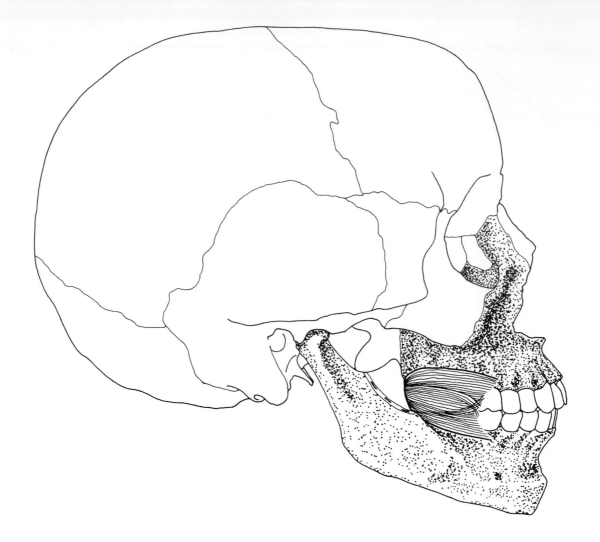

Skull—lateral view

■ Origin
Outer surface of alveolar processes of maxilla and mandible over molars and along pterygomandibular raphe

■ Insertion
Deep part of muscles of lips

■ Action
Compresses cheek

■ Nerve
Buccal branches of facial nerve

TEMPORALIS

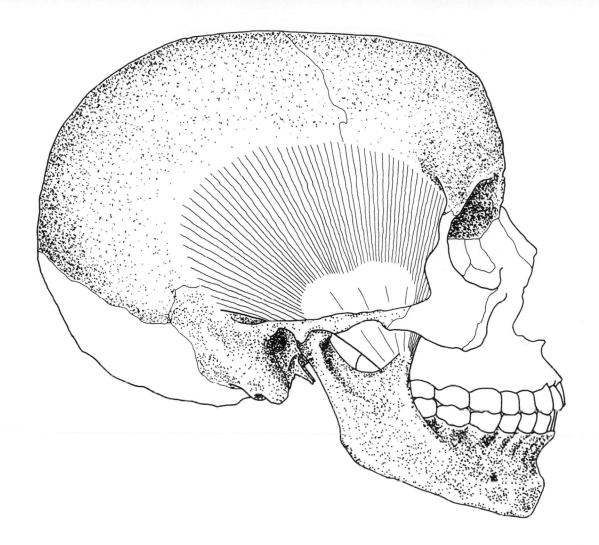

Skull—lateral view

■ Origin
Temporal fossa including frontal, parietal, and temporal bones

■ Insertion
Coronoid process and anterior border of ramus of mandible

■ Action
Elevates, retracts mandible (rotation at tempero-mandibular joint) closing mouth, biting

■ Nerve
Mandibular division of trigeminal nerve

MASSETER

Skull—lateral view

■ Origin
Zygomatic process of maxilla, medial and inferior surfaces of zygomatic arch

■ Insertion
Angle and ramus of mandible, lateral surface of coronoid process of mandible

■ Action
Elevates mandible (rotation at temperomandibular joint) closing mouth, biting

■ Nerve
Mandibular division of trigeminal nerve

Note: Superficial fibers slightly protract jaw (see lateral pterygoid).

MEDIAL PTERYGOID *(Pterygoideus Medialis)*

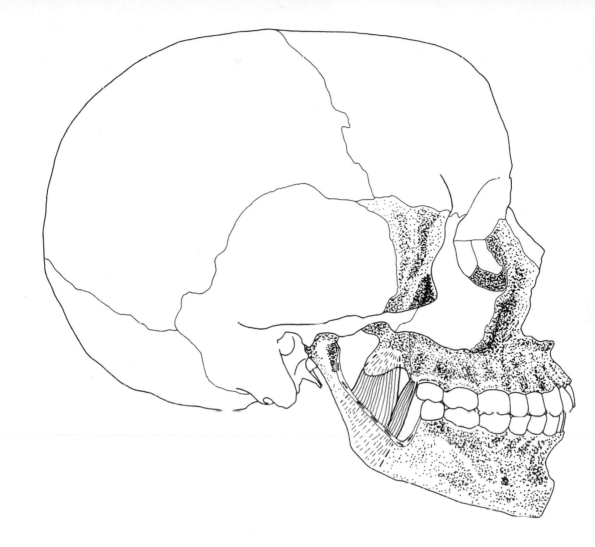

Skull—lateral view
(Part of mandible cut away)

■ **Origin**
Medial surface of lateral pterygoid plate of sphenoid bone, palatine bone, and tuberosity of maxilla

■ **Insertion**
Medial surface of ramus and angle of mandible

■ **Action**
Elevates mandible, assists in protruding mandible

■ **Nerve**
Mandibular division of trigeminal nerve

LATERAL PTERYGOID *(Pterygoideus Lateralis)*

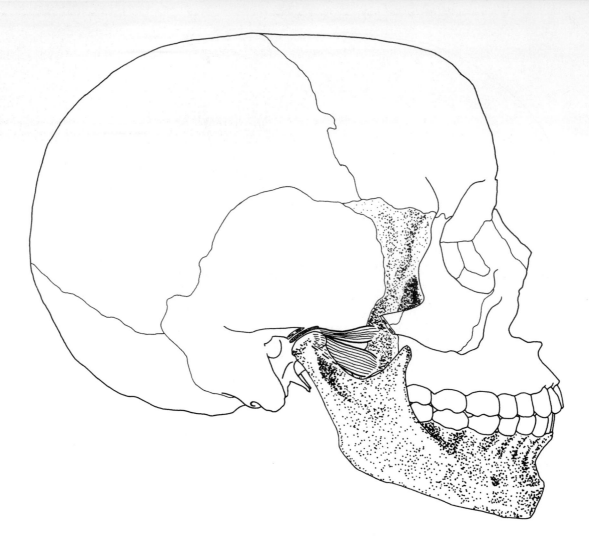

Skull—lateral view

■ Origin
Superior head—lateral surface of greater wing of sphenoid

Inferior head—lateral surface of lateral pterygoid plate

■ Insertion
Condyle of mandible, temporomandibular joint

■ Action
Opens jaw, protrudes mandible, moves mandible laterally for grinding teeth

■ Nerve
Mandibular division of trigeminal nerve

Note: This sideward movement, aided by superficial fibers of masseter, causes chewing movements.

Muscles of
the Neck

STERNOCLEIDOMASTOID

Three-quarter frontal view

■ Origin
Sternal head—manubrium of sternum

Clavicular head—medial part of clavicle

■ Insertion
Mastoid process of temporal bone, lateral half of superior nuchal line of occipital bone

■ Action
Unilateral—bends neck laterally, rotates head to opposite side

Bilateral—flexes neck, draws head ventrally and elevates chin, draws sternum superiorly in deep inspiration

■ Nerve
Spinal part of accessory nerve (C2, C3)

PLATYSMA

Three-quarter frontal view

■ Origin
Subcutaneous fascia of upper one-fourth of chest just below the clavicle

■ Insertion
Subcutaneous fascia and muscles of chin and jaw, mandible

■ Action
Depresses and draws lower lip laterally, draws up skin of chest, depresses mandible

■ Nerve
Cervical branch of facial nerve

DIGASTRICUS

Lateral view

■ Origin

Posterior belly—mastoid notch of temporal bone

Anterior belly—inner side of inferior border of mandible near symphysis

■ Insertion

Intermediate tendon attached to hyoid bone

■ Action

Raises hyoid bone, depresses mandible, moves hyoid forward or backward

■ Nerve

Posterior belly—facial nerve

Anterior belly—mandibular division of trigeminal

Note: The two bellies are joined at an intermediate tendon by a fibrous loop at the side of the body and the greater horn of the hyoid bone.

STYLOHYOID

Lateral view

■ **Origin**
Styloid process of temporal bone

■ **Insertion**
Hyoid bone

■ **Action**
Elevates and draws hyoid bone posteriorly, elevates tongue

■ **Nerve**
Facial nerve

MYLOHYOID

Lateral view

■ Origin
Inside surface of mandible from symphysis to molars (mylohyoid line)

■ Insertion
Hyoid bone

■ Action
Elevates hyoid bone, raises floor of mouth and tongue

■ Nerve
Mandibular division of trigeminal nerve

GENIOHYOID

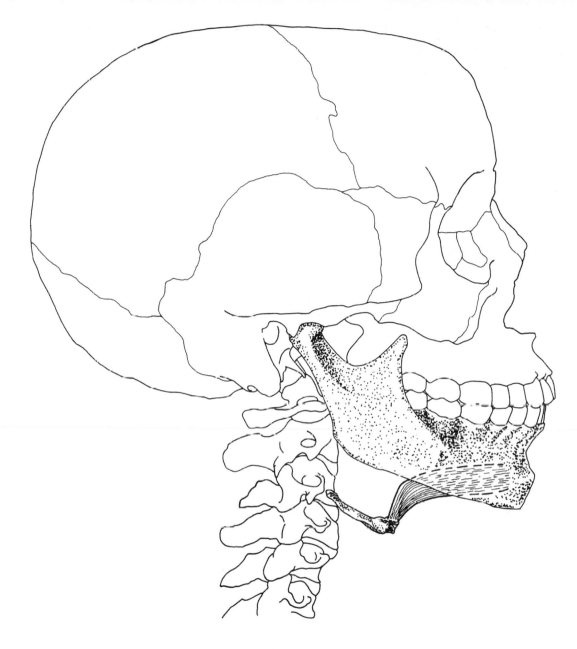

Lateral view

■ **Origin**
Inferior mental spine on interior medial surface of mandible

■ **Insertion**
Body of hyoid bone

■ **Action**
Protrudes hyoid bone and tongue

■ **Nerve**
Branch of C1 through hypoglossal nerve

STERNOHYOID

Frontal view

■ **Origin**
Medial end of clavicle, manubrium of sternum

■ **Insertion**
Body of hyoid bone

■ **Action**
Depresses hyoid bone

■ **Nerve**
Ansa cervicalis (C1–C3)

Lateral view

STERNOTHYROID

Frontal view

■ **Origin**
Dorsal surface of manubrium of sternum

■ **Insertion**
Lamina of thyroid cartilage

■ **Action**
Depresses thyroid cartilage

■ **Nerve**
Ansa cervicalis (C1–C3)

Lateral view

THYROHYOID

Lateral view

■ **Origin**
Lamina of thyroid cartilage

■ **Insertion**
Inferior border of body and greater horn of hyoid bone

■ **Action**
Depresses hyoid or raises thyroid

■ **Nerve**
C1 through hypoglossal nerve

OMOHYOID

Lateral view

■ **Origin**
Superior border of scapula

■ **Insertion**
Inferior belly—bound to clavicle by central tendon

Superior belly—continues to body of hyoid bone

■ **Action**
Depresses hyoid bone

■ **Nerve**
Ansa cervicalis (C2, C3)

LONGUS COLLI

Frontal view
(Mandible and part of maxilla removed)

Superior oblique part

■ **Origin**

Transverse processes of third, fourth, and fifth cervical vertebrae

■ **Insertion**

Anterior arch of atlas

Inferior oblique part

■ **Origin**

Anterior surface of bodies of first two or three thoracic vertebrae

■ **Insertion**

Transverse processes of fifth and sixth cervical vertebrae

Vertical part

■ **Origin**

Anterior surfaces of bodies of upper three thoracic and lower three cervical vertebrae

■ **Insertion**

Anterior surfaces of the second, third, and fourth cervical vertebrae

■ **Action**

All three parts flex cervical vertebrae

■ **Nerve**

C2–C7

Note: Cervical hyperextension injuries (whiplash) may strain these muscles and/or sprain the anterior ligaments of vertebrae.

LONGUS CAPITIS

Frontal view
(Mandible and part of maxilla removed)

■ Origin
Anterior tubercles of transverse processes of third through sixth cervical vertebrae

■ Insertion
Occipital bone anterior to foramen magnum

■ Action
Acting together (bilaterally)—flex head

Acting on one side only—rotate head

■ Nerve
C1–C3

RECTUS CAPITIS ANTERIOR

Frontal view
(Mandible and part of maxilla removed)

■ **Origin**
Anterior base of transverse process of atlas

■ **Insertion**
Occipital bone anterior to foramen magnum

■ **Action**
Flexes head

■ **Nerve**
C1, C2

RECTUS CAPITIS LATERALIS

Frontal view
(Mandible and part of maxilla removed)

■ **Origin**
Transverse process of atlas

■ **Insertion**
Jugular process of occipital bone

■ **Action**
Bends head laterally

■ **Nerve**
C1, C2

SCALENUS ANTERIOR

Three-quarter frontal view
(Mandible and part of maxilla removed)

■ **Origin**
Transverse processes of third through sixth cervical vertebrae

■ **Insertion**
Inner border of first rib (scalene tubercle)

■ **Action**
Raises first rib (respiratory inspiration); acting together, they flex neck; acting on one side, they laterally flex, rotate neck

■ **Nerve**
Ventral rami of cervical nerves (C4–C6)

SCALENUS MEDIUS

Three-quarter frontal view
(Mandible and part of maxilla removed)

■ Origin
Transverse processes of lower six cervical vertebrae
(C2–C7)

■ Insertion
Upper surface of first rib

■ Action
Raises first rib (respiratory inspiration); acting together,
they flex neck; acting on one side, they laterally flex,
rotate neck

■ Nerve
Ventral rami of cervical nerves (C3–C8)

SCALENUS POSTERIOR

Three-quarter frontal view
(Mandible and part of maxilla removed)

■ **Origin**
Transverse processes of lower two or three cervical vertebrae (C5–C7)

■ **Insertion**
Outer surface of second rib

■ **Action**
Raises second rib (respiratory inspiration); acting together, they flex neck; acting on one side, they laterally flex, rotate neck

■ **Nerve**
Ventral rami of lower three cervical nerves

RECTUS CAPITIS POSTERIOR MAJOR

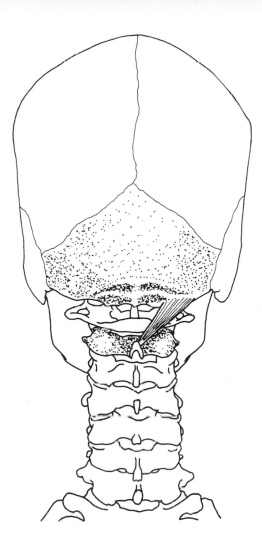

Posterior skull and cervical vertebrae

■ **Origin**
Spinous process of axis

■ **Insertion**
Lateral portion of inferior nuchal line of occipital bone

■ **Action**
Extends and rotates head

■ **Nerve**
Suboccipital nerve

RECTUS CAPITIS POSTERIOR MINOR

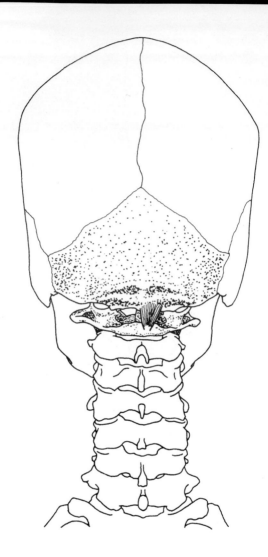

Posterior skull and cervical vertebrae

■ **Origin**
Posterior arch of atlas

■ **Insertion**
Medial portion of inferior nuchal line of occipital bone

■ **Action**
Extends head

■ **Nerve**
Suboccipital nerve

OBLIQUUS CAPITIS INFERIOR

Posterior skull and cervical vertebrae

■ **Origin**
Spinous process of axis

■ **Insertion**
Transverse process of atlas

■ **Action**
Rotates atlas

■ **Nerve**
Suboccipital nerve

OBLIQUUS CAPITIS SUPERIOR

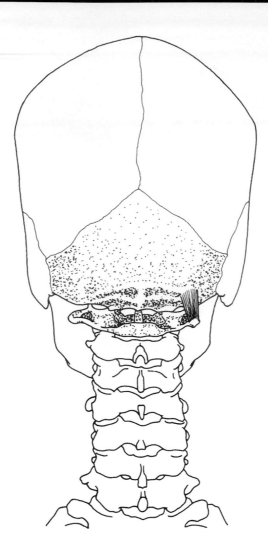

Posterior skull and cervical vertebrae

■ **Origin**
Transverse process of atlas

■ **Insertion**
Occipital bone between inferior and superior nuchal lines

■ **Action**
Extends head and flexes head laterally to the same side (ipsilaterally)

■ **Nerve**
Suboccipital nerve

CHAPTER FIVE

Muscles of the Trunk

SPLENIUS CAPITIS

Posterior skull, neck, and back

■ Origin
Lower part of ligamentum nuchae, spinous processes of seventh cervical vertebra (C7) and upper three or four thoracic vertebrae (T1–T4)

■ Insertion
Mastoid process of temporal bone and lateral part of superior nuchal line

■ Action
Acting together, they extend, hyperextend head, neck; acting on one side, they laterally flex, rotate head, neck

■ Nerve
Lateral branches of dorsal primary divisions of middle and lower cervical nerves

SPLENIUS CERVICIS

Posterior skull, neck, and back

■ Origin
Spinous processes of third through sixth thoracic vertebrae (T3–T6)

■ Insertion
Transverse processes of upper two or three cervical vertebrae (C1–C3)

■ Action
Acting together, they extend, hyperextend head, neck; acting on one side, they laterally flex, rotate head, neck

■ Nerve
Lateral branches of dorsal primary divisions of middle and lower cervical nerves

ERECTOR SPINAE

Iliocostalis cervicis

Origin
Angles of third through sixth ribs

Insertion
Transverse processes of fourth, fifth, and sixth cervical vertebrae

Action
Extension, lateral flexion of vertebral column

Nerve
Dorsal primary divisions of spinal nerves

Iliocostalis thoracis

Origin
Angles of lower six ribs medial to iliocostalis lumborum

Insertion
Angles of upper six ribs and transverse process of seventh cervical vertebra

Action
Extension, lateral flexion of vertebral column, rotates ribs for forceful inspiration

Nerve
Dorsal primary divisions of spinal nerves

Iliocostalis lumborum

Origin
Medial and lateral sacral crests and medial part of iliac crests

Insertion
Angles of lower six ribs

Action
Extension, lateral flexion of vertebral column, rotates ribs for forceful inspiration

Nerve
Dorsal primary divisions of spinal nerves

Note: The erector spinae (sacrospinalis) is a complex of three sets of muscles: iliocostalis, longissimus, and spinalis. The origin of this group is the medial and lateral sacral crests, the medial part of iliac crests, and the spinous processes and supraspinal ligament of lumbar and eleventh and twelfth thoracic vertebrae.

Trunk—dorsal view

ERECTOR SPINAE

Trunk—dorsal view

Longissimus capitis

■ Origin
Transverse processes of upper five thoracic vertebrae (T1–T5), articular processes of lower three cervical vertebrae (C5–C7)

■ Insertion
Posterior part of mastoid process of temporal bone

■ Action
Extends and rotates head

■ Nerve
Dorsal primary divisions of middle and lower cervical nerves

Longissimus cervicis

■ Origin
Transverse processes of upper four or five thoracic vertebrae (T1–T5)

■ Insertion
Transverse processes of second through sixth cervical vertebrae (C2–C6)

■ Action
Extension, lateral flexion of vertebral column

■ Nerve
Dorsal primary divisions of spinal nerves

Longissimus thoracis

■ Origin
Medial and lateral sacral crests, spinous processes and supraspinal ligament of lumbar and eleventh and twelfth thoracic vertebrae, and medial part of iliac crests

■ Insertion
Transverse processes of all thoracic vertebrae, between tubercles and angles of lower nine or ten ribs

■ Action
Extension, lateral flexion of vertebral column, rotates ribs for forceful inspiration

■ Nerve
Dorsal primary divisions of spinal nerves

ERECTOR SPINAE

Spinalis capitis
(Medial part of semispinalis capitis)

Spinalis cervicis

Origin
Ligamentum nuchae, spinous process of seventh cervical vertebra

Insertion
Spinous process of axis

Action
Extends vertebral column

Nerve
Dorsal primary divisions of spinal nerves

Spinalis thoracis

Origin
Spinous processes of lower two thoracic (T11, T12) and upper two lumbar (L1, L2) vertebrae

Insertion
Spinous processes of upper thoracic vertebrae (T1–T8)

Action
Extends vertebral column

Nerve
Dorsal primary divisions of spinal nerves

Trunk—dorsal view

TRANSVERSOSPINALIS

Trunk—dorsal view

Semispinalis capitis
(Medial part is spinalis capitis)

■ Origin
Articular processes of cervical vertebrae C4–C6 and transverse processes of C7 and upper six or seven thoracic vertebrae (T1–T7)

■ Insertion
Between superior and inferior nuchal lines of occipital bone

■ Action
Extends and rotates head

■ Nerve
Dorsal primary divisions of spinal nerves

Semispinalis cervicis

■ Origin
Transverse processes of upper five or six thoracic vertebrae (T1–T6)

■ Insertion
Spinous processes of second to fifth cervical vertebrae (C2–C5)

■ Action
Extends and rotates vertebral column

■ Nerve
Dorsal primary divisions of spinal nerves

Semispinalis thoracis

■ Origin
Transverse processes of the sixth through tenth thoracic vertebrae (T6–T10)

■ Insertion
Spinous processes of the lower two cervical (C6, C7) and upper four thoracic (T1–T4) vertebrae

■ Action
Extends and rotates vertebral column

■ Nerve
Dorsal primary divisions of spinal nerves

Note: The transversospinalis is composed of groups of small muscles generally extending upward from transverse processes to spinous processes of higher vertebrae. They are deep to erector spinae. They include semispinalis, multifidi, and rotatores.

MULTIFIDIS*

Trunk—dorsal view

■ Origin

Sacral region—along sacral foramina up to posterior superior iliac spine

Lumbar region—mammillary processes† of vertebrae

Thoracic region—transverse processes

Cervical region—articular processes of lower four vertebrae (C4–C7)

■ Insertion

Spinous process two to four vertebrae superior to origin

■ Action

Extend and rotate vertebral column

■ Nerve

Dorsal primary division of spinal nerves

*Part of transversospinalis.

†Posterior border of superior articular process.

ROTATORES*

Trunk—dorsal view

■ Origin
Transverse process of each vertebra

■ Insertion
Base of spinous process of next vertebra above

■ Action
Extend and rotate vertebral column

■ Nerve
Dorsal primary division of spinal nerves

*Part of transversospinalis.

INTERSPINALES *(Paired on either side of interspinal ligament)*

Trunk—dorsal view

■ Origin

Cervical region—spinous processes of third to seventh cervical vertebrae (C3–C7)

Thoracic region—spinous processes of first to third (T1–T3) and eleventh and twelfth thoracic vertebrae (T11, T12)

Lumbar region—spinous processes of second to fifth lumbar vertebrae (L2–L5)

■ Insertion

Spinous process of next vertebra superior to origin

■ Action

Extend vertebral column

■ Nerve

Dorsal primary division of spinal nerves

INTERTRANSVERSARII

Trunk—dorsal view

Cervical region

Intertransversarii anteriores

- **Origin**   Anterior tubercle of transverse processes of vertebrae from first thoracic to axis
- **Insertion**   Anterior tubercle of next superior vertebra
- **Action**   Lateral flexion of vertebral column
- **Nerve**   Ventral primary division of spinal nerves

Intertransversarii posteriores

- **Origin**   Posterior tubercle of transverse processes of vertebrae from first thoracic to axis
- **Insertion**   Posterior tubercle of next superior vertebra

Thoracic region

- **Origin**   Transverse processes of first lumbar to eleventh thoracic vertebrae
- **Insertion**   Transverse processes of next superior vertebra

Lumbar region

Intertransversarii laterales

- **Origin**   Transverse processes of lumbar vertebrae
- **Insertion**   Transverse process of next superior vertebra
- **Action**   Lateral flexion of vertebral column
- **Nerve**   Ventral primary division of spinal nerves

Intertransversarii mediales

- **Origin**   Mammillary process* of each lumbar vertebra
- **Insertion**   Accessory process of the next superior lumbar vertebra
- **Action**   Lateral flexion of vertebral column
- **Nerve**   Dorsal primary division of spinal nerves

*Posterior border of superior articular process.

EXTERNAL INTERCOSTALS *(Intercostales Externi)*

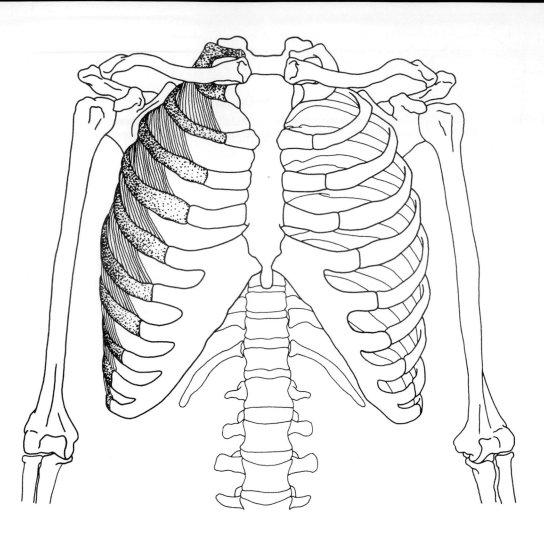

Trunk—anterior view

■ Origin
Lower margin of upper eleven ribs

■ Insertion
Superior border of rib below (each muscle fiber runs obliquely and inserts toward the costal cartilage)

■ Action
Draw ventral part of ribs upward, increasing the volume of the thoracic cavity for inspiration

■ Nerve
Intercostal nerves

INTERNAL INTERCOSTALS *(Intercostales Interni)*

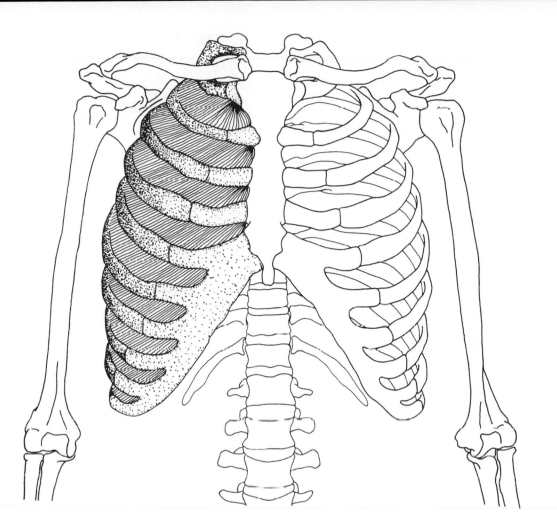

Trunk—anterior view

■ Origin
From the cartilages to the angles of the upper eleven ribs

■ Insertion
Superior border of the rib below (each muscle fiber runs obliquely and inserts away from the costal cartilage)

■ Action
Draw ventral part of ribs downward, decreasing the volume of the thoracic cavity for expiration

■ Nerve
Intercostal nerves

SUBCOSTALS *(Subcostales)*

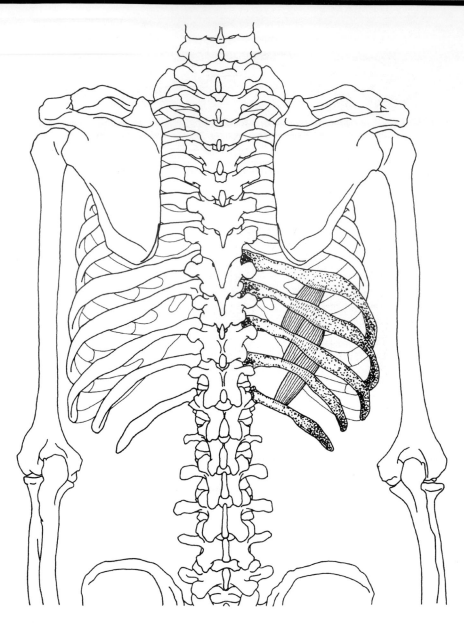

Trunk—dorsal view

■ Origin
Inner surface of each rib near its angle

■ Insertion
Medially on the inner surface of second or third rib below

■ Action
Draw ventral part of ribs downward, decreasing the volume of the thoracic cavity for forceful expiration

■ Nerve
Intercostal nerves

Note: These muscles are deep to the internal intercostals. They continue distally between single ribs, where they are known as innermost intercostal muscles.

TRANSVERSUS THORACIS

Trunk—anterior view

■ Origin
Inner surface of lower portion of sternum and adjacent costal cartilages

■ Insertion
Inner surfaces of costal cartilages of the second through sixth ribs

■ Action
Draws ventral part of ribs downward, decreasing the volume of the thoracic cavity for forceful expiration

■ Nerve
Intercostal nerves

Note: These muscles are deep to the internal intercostal muscles.

LEVATORES COSTARUM

Trunk—dorsal view

■ Origin
Transverse processes of the seventh cervical and the upper eleven thoracic vertebrae

■ Insertion
Laterally to outer surface of next lower rib (lower muscles may cross over one rib)

■ Action
Raises ribs; extends, laterally flexes, and rotates vertebral column

■ Nerve
Intercostal nerves

SERRATUS POSTERIOR SUPERIOR

Trunk—dorsal view

■ **Origin**
Ligamentum nuchae, spinous processes of seventh cervical and first few thoracic vertebrae

■ **Insertion**
Upper borders of the second through fifth ribs lateral to their angles

■ **Action**
Raises ribs in inspiration

■ **Nerve**
T1–T4

SERRATUS POSTERIOR INFERIOR

Trunk—dorsal view

■ Origin
Spinous processes of the lower two thoracic and the upper two or three lumbar vertebrae

■ Insertion
Lower borders of bottom four ribs

■ Action
Pulls ribs down, resisting pull of diaphragm

■ Nerve
T9–T12

DIAPHRAGM

Trunk—anterior view
(Lower costal cartilages removed)

■ Origin

Sternal part—inner part of xiphoid process

Costal part—inner surfaces of lower six ribs and their cartilages

Lumbar part—upper two or three lumbar vertebrae and lateral and medial lumbocostal arches*

■ Insertion

Fibers converge and meet on a central tendon

■ Action

Draws central tendon inferiorly, for inspiration

■ Nerve

Phrenic nerve (C3–C5)

Note: This muscle inserts upon itself. Its action is to change the volume of the thoracic and abdominal cavities.

*These tendinous structures, also known as the medial and lateral arcuate ligaments, allow the diaphragm to bridge the upper parts of the psoas major and quadratus lumborum muscles.

EXTERNAL OBLIQUE *(Obliquus Externus Abdominis)*

Trunk—lateral view

■ Origin
Lower eight ribs

■ Insertion
Anterior part of iliac crest, abdominal aponeurosis (rectus sheath) to linea alba

■ Action
Compresses abdominal contents, laterally flexes and rotates vertebral column

■ Nerve
Eighth to twelfth intercostal, iliohypogastric, ilioinguinal nerves

■ Relationship
Most superficial of the three lateral abdominal muscles

Note: Important in forced expiration, coughing, sneezing.

INTERNAL OBLIQUE *(Obliquus Internus Abdominis)*

Trunk—lateral view

■ Origin
Lateral half of inguinal ligament, iliac crest, thoracolumbar fascia

■ Insertion
Cartilage of bottom three or four ribs, abdominal aponeurosis (rectus sheath) to linea alba

■ Action
Compresses abdominal contents, laterally flexes and rotates vertebral column

■ Nerve
Eighth to twelfth intercostal, iliohypogastric, ilioinguinal nerves

■ Relationship
Middle layer of the three lateral abdominal muscles

Note: Important in forced expiration, coughing, sneezing.

CREMASTER

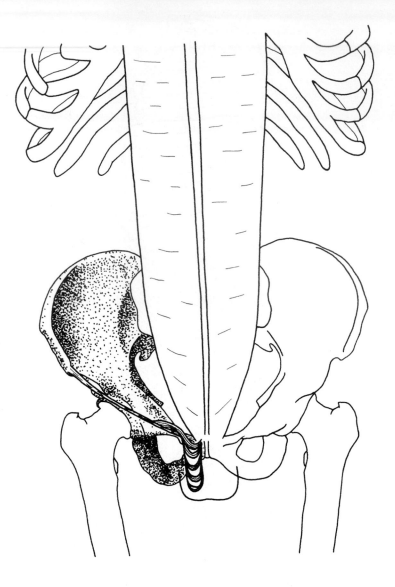

Trunk—anterior view

■ Origin
Inguinal ligament

■ Insertion
Pubic tubercle, crest of pubis, sheath of rectus abdominis

■ Action
Pulls testes toward body

■ Nerve
Genital branch of genitofemoral nerve

■ Relationship
This muscle is continuous with the internal abdominal oblique.

Note: The cremaster regulates the temperature of the testes, important for spermatogenesis.

TRANSVERSE ABDOMINAL *(Transversus Abdominis)*

Trunk—lateral view

■ Origin
Lateral part of inguinal ligament, iliac crest, thoracolumbar fascia, cartilage of lower six ribs

■ Insertion
Abdominal aponeurosis (rectus sheath) to linea alba

■ Action
Compresses abdomen

■ Nerve
Seventh to twelfth intercostal, iliohypogastric, ilioinguinal nerves

■ Relationship
Deepest of the three lateral abdominal muscles

Note: Important in forced expiration, coughing, sneezing.

RECTUS ABDOMINIS*

Trunk—anterior view

■ Origin
Crest of pubis, pubic symphysis

■ Insertion
Cartilage of fifth, sixth, and seventh ribs, xiphoid process

■ Action
Flexes vertebral column, compresses abdomen

■ Nerve
Seventh through twelfth intercostal nerves

*Tendinous bands divide each rectus into three or four bellies. Each rectus is sheathed in aponeurotic fibers from the lateral abdominal muscles. These fibers meet centrally to form the linea alba.

Note: The pyramidalis is a small, unimportant muscle that extends from the ventral surface of the pubis to the lower part of the linea alba. It is frequently absent.

ABDOMINAL MUSCLES

Trunk—anterior and cross-sectional views

1. External oblique
2. Internal oblique
3. Transverse abdominal

4. Rectus abdominis
5. Linea alba

Note: The aponeuroses (broad, flat tendons) of the three lateral abdominal muscles join to form the fascial sheath surrounding the rectus abdominis.

The rectus abdominis is separated into four bellies by tendinous sheaths.

QUADRATUS LUMBORUM

Lower trunk—anterior view

■ **Origin**
Iliolumbar ligament, iliac crest

■ **Insertion**
Twelfth rib, transverse processes of upper four lumbar vertebrae

■ **Action**
Laterally flexes vertebral column, fixes ribs for forced expiration

■ **Nerve**
T12, L1

Note: Fixation of the ribs may provide a stable attachment of the diaphragm for voice control in singers.

CHAPTER SIX

Muscles of the Shoulder and Arm

PECTORALIS MAJOR

Anterior view

■ Origin
Clavicular part—medial half of the clavicle

Sternocostal part—sternum, upper six costal cartilages, aponeurosis of external oblique

■ Insertion
Lateral lip of intertubercular (bicipital) groove of humerus, crest below greater tubercle of the humerus

■ Action
Both parts adduct, medially rotate arm; clavicular part flexes arm from full extension; sternocostal part extends the flexed arm

■ Nerve
Medial and lateral pectoral nerves (C5–C8, T1)

PECTORALIS MINOR

Anterior view

■ Origin
External surfaces of the third, fourth, and fifth ribs

■ Insertion
Coracoid process of the scapula

■ Action
Draws scapula forward and downward, raises ribs* in forced inspiration

■ Nerve
Medial pectoral nerve (C8, T1)

■ Relationship
Deep to pectoralis major; medial pectoral nerve pierces this muscle

*Raising the ribs requires stabilization of the scapula by the rhombolds and trapezius.

SUBCLAVIUS

Anterior view

■ Origin
Junction of the first rib with its costal cartilage

■ Insertion
Groove on the inferior (lower) surface of the clavicle

■ Action
Steadies clavicle during movements of the shoulder girdle

■ Nerve
Nerve to the subclavius from upper trunk of brachial plexus (C5, C6)

CORACOBRACHIALIS

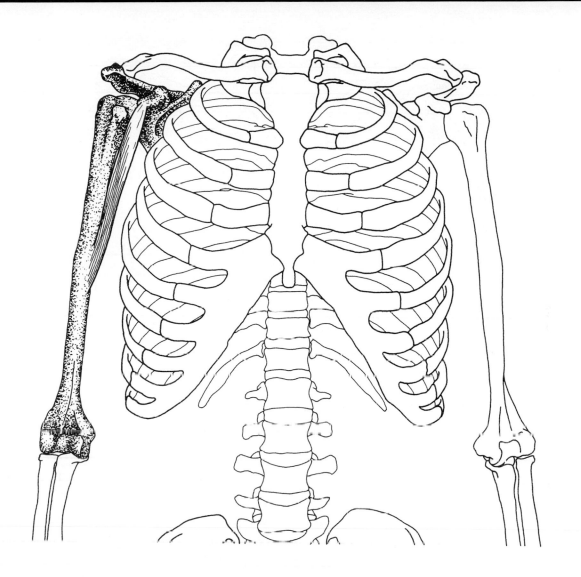

Anterior view

■ **Origin**
Tip (apex) of the coracoid process of scapula

■ **Insertion**
Middle third of the medial surface and border of the humerus

■ **Action**
Weakly adducts arm (flexion unsubstantiated), aids in stabilizing humerus

■ **Nerve**
Musculocutaneous nerve (C6, C7)

■ **Relationship**
Deep to short head of biceps

Note: The musculocutaneous nerve usually pierces this muscle.

BICEPS BRACHII

Anterior view

■ Origin

Long head—supraglenoid tubercle of scapula

Short head—coracoid process of scapula

■ Insertion

Tuberosity of radius, bicipital aponeurosis into deep fascia on medial part of forearm

■ Action

Supinates forearm, flexes forearm, weakly flexes arm at shoulder

■ Nerve

Musculocutaneous nerve (C5, C6)

■ Relationship

Long head passes through intertubercular (bicipital) groove, then inside glenohumeral joint capsule

Note: As a two joint muscle, its contribution to shoulder flexion mainly occurs when both joints are extended.

BRACHIALIS

Anterior view

■ **Origin**
Anterior of lower half of humerus

■ **Insertion**
Coronoid process of ulna, tuberosity of ulna

■ **Action**
Flexes forearm

■ **Nerve**
Musculocutaneous nerve (C5, C6)

■ **Relationship**
Deep to biceps brachii

MUSCLES OF THE ANTERIOR CHEST AND ARM

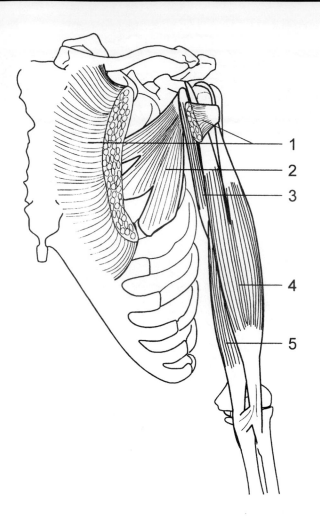

Shoulder—anterior view

1. Pectoralis major (cut)
2. Pectoralis minor
3. Coracobrachialis
4. Biceps brachii
5. Brachialis

TRAPEZIUS

Posterior view

■ Origin
Medial third of superior nuchal line, external occipital protuberance, ligamentum nuchae, spinous processes and supraspinous ligaments of seventh cervical and all thoracic vertebrae

■ Insertion
Upper part—lateral third of clavicle

Middle part—acromion and spine of scapula

Lower part—medial portion of spine of scapula (tubercle)

■ Action
Upper part elevates scapula*, middle part retracts (adducts) scapula, lower part depresses scapula, upper and lower parts together rotate scapula (important in elevating arm)

■ Nerve
Accessory (eleventh cranial), C3, C4

■ Relationship
Most superficial muscle of back

*Upper part stabilizes scapula against downward rotation, as when weight is carried in the hand.

LATISSIMUS DORSI

Posterior view

■ Origin
Spinous processes of lower six thoracic vertebrae, spinous processes of all lumbar and sacral vertebrae (through thoracolumbar fascia), supraspinal ligament and iliac crest, outer surfaces of lower three or four ribs, inferior angle of scapula

■ Insertion
Bottom of intertubercular (bicipital) groove

■ Action
Extends, adducts, and medially rotates the arm; draws the shoulder downward and backward; keeps inferior angle of scapula against the chest wall; accessory muscle of respiration

■ Nerve
Thoracodorsal nerve (C6–C8)

Note: This muscle is used for the crawl (freestyle) stroke in swimming.

LEVATOR SCAPULAE

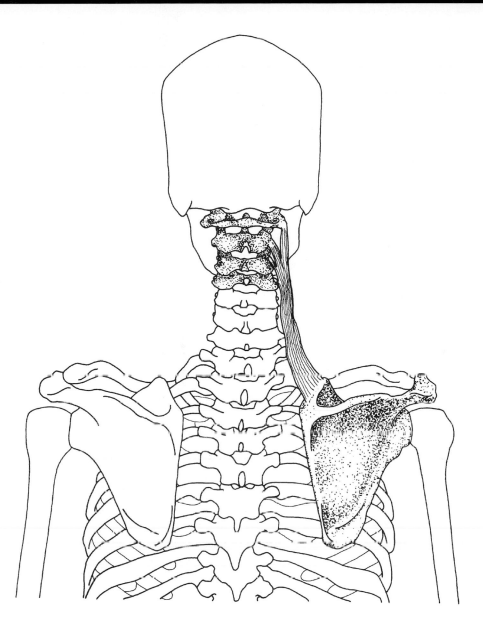

Posterior view

■ Origin
Posterior tubercles of the transverse processes of the first four cervical vertebrae

■ Insertion
Vertebral (medial) border of the scapula at and above the spine

■ Action
Elevates medial border of scapula, rotates scapula to lower the lateral angle, acts with trapezius and rhomboids to pull scapula medially and upward, bends neck laterally

■ Nerve
Dorsal scapular nerve (C5)

RHOMBOID MAJOR

Posterior view

■ Origin
Spines of the second to fifth thoracic vertebrae, supraspinous ligament

■ Insertion
Medial border of the scapula below the spine

■ Action
Retracts and stabilizes scapula, elevates the medial border of the scapula causing downward rotation, assists in adduction of arm

■ Nerve
Dorsal scapular nerve (C5)

RHOMBOID MINOR

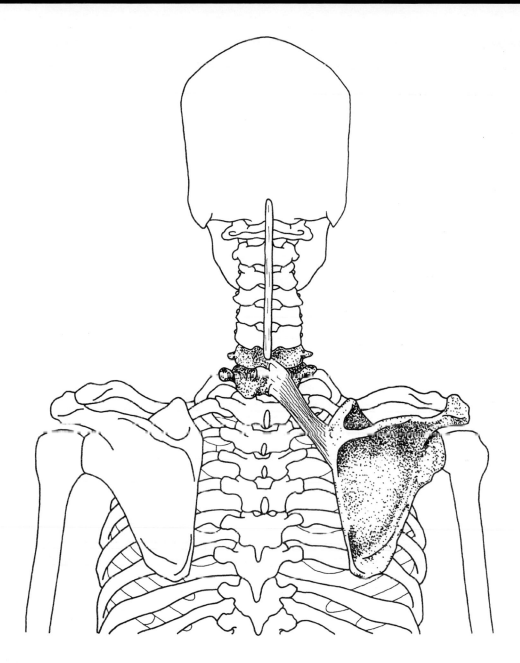

Posterior view

■ Origin
Spines of the seventh cervical and first thoracic vertebrae, lower part of the ligamentum nuchae

■ Insertion
Medial border of the scapula at the root of the spine

■ Action
Retracts and stabilizes scapula, elevates the medial border of the scapula, rotates the scapula to depress the lateral angle (assists in adduction of arm)

■ Nerve
Dorsal scapular nerve (C5)

SERRATUS ANTERIOR

Lateral view

■ Origin
Outer surfaces and superior borders of first eight or nine ribs, and fascia covering first intercostal space

■ Insertion
Anterior surface (costal surface) of the medial border of the scapula

■ Action
Rotates scapula for abduction and flexion of arm, protracts scapula

■ Nerve
Long thoracic nerve (C5–C7)

■ Relationships
Serratus anterior and rhomboids both insert on the medial border of scapula; they are antagonists causing protraction and retraction; long thoracic nerve lies on the superficial surface of this muscle

Note: When the arm is fixed, this muscle can assist in rib movement for ventilation.

DELTOID

Lateral view

■ Origin
Anterior portion—anterior border and superior surface of the lateral third of the clavicle

Middle portion—lateral border of the acromion process

Posterior portion—lower border of the crest of the spine of the scapula

■ Insertion
Deltoid tuberosity, on the middle of the lateral surface of the shaft of the humerus

■ Action
Anterior portion—flexes and medially rotates arm

Middle portion—abducts arm

Posterior portion—extends and laterally rotates arm

■ Nerve
Axillary nerve (C5, C6)

SUPRASPINATUS *(Rotator Cuff*)*

Lateral view

■ Origin
Supraspinous fossa of scapula

■ Insertion
Upper part of the greater tuberosity of the humerus, capsule of the shoulder joint

■ Action
Aids deltoid in abduction of arm; draws humerus toward glenoid fossa, preventing deltoid from forcing humerus up against acromion

■ Nerve
Suprascapular nerve (C5)

*Supraspinatus, infraspinatus, teres minor, and subscapularis together are called the rotator cuff.

Note: The joint capsule and its ligaments cannot provide necessary support because of the great range of motion between the humerus and scapula. The rotator cuff muscles prevent dislocation of the humerus throughout most of the arm's range of motion.

INFRASPINATUS *(Rotator Cuff)*

Lateral view

■ Origin
Infraspinous fossa of the scapula

■ Insertion
Middle facet of the greater tuberosity of the humerus, capsule of the shoulder joint

■ Action
Draws humerus toward glenoid fossa, thus resisting posterior dislocation of arm, as in crawling; laterally rotates; aids in stabilization of humerus during abduction

■ Nerve
Suprascapular nerve (C5, C6)

TERES MINOR *(Rotator Cuff)*

Lateral view

■ Origin
Upper two-thirds of the dorsal surface of the axillary border of the scapula

■ Insertion
The capsule of the shoulder joint, the lower facet of the greater tuberosity of the humerus

■ Action
Laterally rotates arm, weakly adducts arm, draws humerus toward glenoid fossa

■ Nerve
Axillary nerve (C5)

SUBSCAPULARIS *(Rotator Cuff)*

Anterior view
(Upper ribs cut away)

■ Origin
Subscapular fossa on the anterior surface of scapula

■ Insertion
Lesser tuberosity (tubercle) of the humerus, ventral part of the capsule of the shoulder joint

■ Action
Medially rotates arm, stabilizes glenohumeral joint

■ Nerve
Upper and lower subscapular nerves (C5, C6)

TERES MAJOR

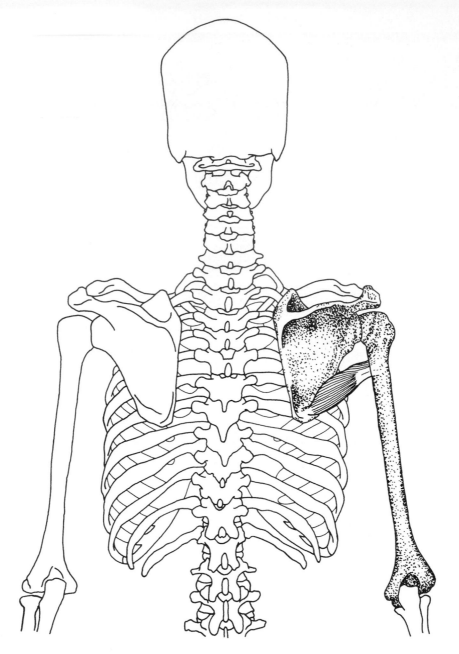

Posterior view

■ Origin
Lower third of the posterior surface of the lateral border of the scapula, near the inferior angle

■ Insertion
Medial lip of the intertubercular (bicipital) groove of the humerus

■ Action
Medially rotates arm, adducts arm, extends arm

■ Nerve
Lower subscapular nerve (C5, C6)

TRICEPS BRACHII

Posterior view

■ Origin

Long head—infraglenoid tubercle of the scapula

Lateral head—upper half of the posterior surface of the shaft of the humerus

Medial head—posterior surface of the lower half of the shaft of the humerus

■ Insertion

Posterior part of olecranon process of the ulna

■ Action

Extends forearm, long head aids in adduction if arm is abducted

■ Nerve

Radial nerve (C7, C8)

Note: The radial nerve comes from the axilla (armpit) and passes along the humerus between the medial and lateral heads. It can be compressed against the humerus, so it is one of the most commonly injured peripheral nerves.

ANCONEUS

Posterior view of arm

■ **Origin**
Posterior part of lateral epicondyle of the humerus

■ **Insertion**
Lateral surface of the olecranon process and posterior surface of ulna

■ **Action**
Extends forearm (assists triceps)

■ **Nerve**
Radial nerve (C7, C8)

POSTERIOR BACK, SHOULDER, AND ARM MUSCLES

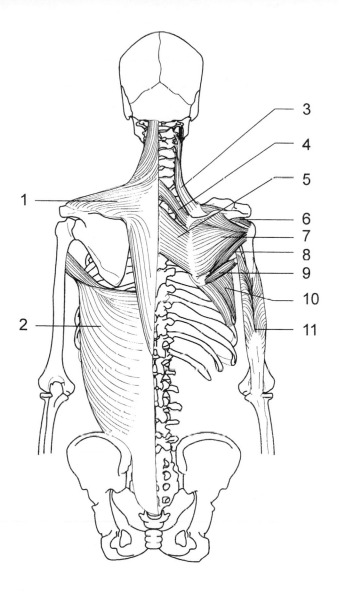

Trunk—posterior view

Superficial layer

1. Trapezius
2. Latissimus dorsi

Deep layer

3. Levator scapulae
4. Rhomboid minor
5. Rhomboid major

6. Supraspinatus (rotator cuff)
7. Infraspinatus (rotator cuff)
8. Teres minor (rotator cuff)
9. Teres major
10. Serratus anterior

Posterior arm

11. Triceps brachii

CHAPTER SEVEN

Muscles of the Forearm and Hand

PRONATOR TERES *(Superficial Group)*

Forearm—anterior view

■ Origin

Humeral head—medial supracondylar ridge and medial epicondyle of the humerus

Ulnar head—medial border of the coronoid process of the ulna

■ Insertion

Middle of lateral surface of the radius (pronator tuberosity)

■ Action

Pronates and flexes forearm

■ Nerve

Median nerve (C6, C7)

Note: The median nerve passes between the humeral and ulnar heads of this muscle.

FLEXOR CARPI RADIALIS *(Superficial Group)*

Forearm—anterior view

■ Origin
Medial epicondyle of the humerus through the common tendon

■ Insertion
Front of the bases of the second and third metacarpal bones

■ Action
Flexes hand, synergist in abduction with extensor carpi radialis longus and brevis

■ Nerve
Median nerve (C6, C7)

PALMARIS LONGUS *(Superficial Group)*

Forearm—anterior view

■ Origin
Medial epicondyle of the humerus through the common tendon

■ Insertion
Front (central part) of the flexor retinaculum and apex of the palmar aponeurosis

■ Action
Flexes the hand

■ Nerve
Median nerve (C6, C7)

Note: This muscle is absent in about 14% of limbs.

Reference: Agur, Amr: *Grant's Atlas of Anatomy,* 9th ed. Williams & Wilkins, Baltimore, 1991.

FLEXOR CARPI ULNARIS *(Superficial Group)*

Forearm—anterior view

■ Origin
Humeral head—medial epicondyle of the humerus through the common tendon

Ulnar head—medial margin of olecranon process of ulna, dorsal border of shaft of the ulna

■ Insertion
Pisiform bone, hook of the hamate, and base of the fifth metacarpal bone

■ Action
Flexes hand, synergist in adduction of hand with extensor carpi ulnaris

■ Nerve
Ulnar nerve (C8, T1)

SUPERFICIAL MUSCLES OF THE FOREARM

Forearm—anterior view

Muscles

1. Pronator teres
2. Flexor carpi radialis
3. Palmaris longus
4. Flexor carpi ulnaris

Associated structures

5. Flexor retinaculum
6. Palmar aponeurosis

FLEXOR DIGITORUM SUPERFICIALIS

Forearm—anterior view

■ Origin
Humeroulnar head—medial epicondyle of the humerus through common tendon*, medial margin of the coronoid process of ulna

Radial head—anterior surface of shaft of radius

■ Insertion
Four tendons divide into two slips each; slips insert into the sides (margins of the anterior surfaces) of the middle phalanges of four fingers

■ Action
Flexes the middle phalanges of the fingers

■ Nerve
Median nerve (C7, C8, T1)

■ Relationship
Deep to superficial flexors

*See superficial flexors.

Note: The tendons of flexor digitorum superficialis split and attach to the middle phalanx. The tendons of flexor digitorum profundus pass through this split and continue to the distal phalanx.

FLEXOR DIGITORUM PROFUNDUS

Forearm—anterior view

■ Origin
Upper three-fourths of anterior and medial surfaces of shaft of ulna and medial side of the coronoid process, interosseous membrane

■ Insertion
Front of base of distal phalanges

■ Action
Flexes distal phalanges

■ Nerve
Ulnar nerve supplies the medial half of the muscle (going to the fourth and fifth fingers)

Anterior interosseous branch of median nerve supplies lateral half (going to index and middle fingers) (C8, T1)

■ Relationship
Deep to flexor digitorum superficialis

Note: Flexor digitorum muscles, flexor pollicis longus, and the median nerve pass under the flexor retinaculum (p. 134) in the wrist. When irritated, the synovial sheaths of these muscles can compress the median nerve, causing the sensory and motor deficits known as carpal tunnel syndrome.

FLEXOR POLLICIS LONGUS

Forearm—anterior view

■ Origin
Middle of anterior surface of shaft of radius, interosseous membrane, medial epicondyle of humerus, and often coronoid process of ulna

■ Insertion
Palmar aspect of base of the distal phalanx of thumb

■ Action
Flexes the thumb

■ Nerve
Anterior interosseous branch of median nerve (C8, T1)

FLEXORS OF THE FINGERS

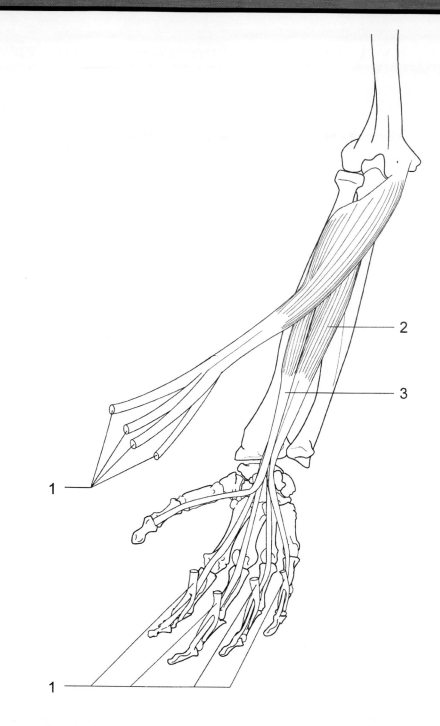

Forearm—anterior view

1. Flexor digitorum superficialis (cut)
2. Flexor digitorum profundus
3. Flexor pollicis longus

PRONATOR QUADRATUS

Forearm—anterior view

■ Origin
Anterior surface of distal part of shaft of ulna

■ Insertion
Lower portion of anterior surface of shaft of radius, distal part of lateral border of radius

■ Action
Pronates forearm and hand

■ Nerve
Anterior interosseous branch of median nerve (C8, T1)

■ Relationship
Deepest forearm muscle

BRACHIORADIALIS

Forearm—dorsal view

■ **Origin**
Upper two-thirds of lateral supracondylar ridge of humerus

■ **Insertion**
Base of styloid process and lateral surface of radius

■ **Action**
Flexes forearm

■ **Nerve**
Radial nerve (C5, C6)

EXTENSOR CARPI RADIALIS LONGUS

Forearm—dorsal view

■ Origin
Lower third of lateral supracondylar ridge of humerus

■ Insertion
Dorsal surface of the base of the second metacarpal bone

■ Action
Extends hand, synergist in abduction of hand with flexor carpi radialis

■ Nerve
Radial nerve (C6, C7)

EXTENSOR CARPI RADIALIS BREVIS

Forearm—dorsal view

■ **Origin**
Lateral epicondyle of humerus

■ **Insertion**
Dorsal surface of third metacarpal bone

■ **Action**
Extends hand, synergist in abduction of hand with flexor carpi radialis

■ **Nerve**
Radial nerve (C7, C8)

EXTENSOR DIGITORUM

Forearm and hand—dorsal view

■ Origin
Common tendon attached to lateral epicondyle of humerus

■ Insertion
Lateral and dorsal surfaces of all the phalanges of the four fingers

■ Action
Extends the fingers and wrist

■ Nerve
Deep branch of radial nerve (C7, C8)

■ Relationship
Tends to hyperextend the metacarpophalangeal joint causing "claw hand"; its action is balanced by the lumbricales and interossei

EXTENSOR DIGITI MINIMI

Forearm and hand—dorsal view

■ **Origin**

Common tendon attached to lateral epicondyle of humerus

■ **Insertion**

Dorsal surface of base of first phalanx of fifth finger

■ **Action**

Extends fifth finger

■ **Nerve**

Radial nerve (C7, C8)

EXTENSOR CARPI ULNARIS

Forearm and hand—dorsal view

■ Origin
Common tendon attached to lateral epicondyle of humerus

■ Insertion
Dorsal surface of base of fifth metacarpal bone

■ Action
Extends hand, synergist in adduction of hand with flexor carpi ulnaris

■ Nerve
Radial nerve (C7, C8)

SUPINATOR

Forearm and hand—anterior view

■ Origin
Lateral epicondyle of humerus, lateral ligament (radial collateral) of elbow, annular ligament of superior radioulnar joint, supinator crest of ulna

■ Insertion
Dorsal and lateral surfaces of upper third of radius

■ Action
Supinates forearm

■ Nerve
Radial nerve (C5, C6)

ABDUCTOR POLLICIS LONGUS

Forearm and hand—dorsal view

■ Origin
Posterior (dorsal) surface of shaft of radius, ulna, interosseous membrane

■ Insertion
Dorsal surface of base of first metacarpal bone

■ Action
Abducts, laterally rotates, and extends thumb; abducts wrist

■ Nerve
Radial nerve (C7, C8)

EXTENSOR POLLICIS BREVIS

Forearm and hand—dorsal view

■ Origin
Dorsal surface of radius, adjacent part of interosseous membrane

■ Insertion
Base of proximal phalanx of thumb

■ Action
Extends thumb, abducts hand

■ Nerve
Radial nerve (C7, C8)

EXTENSOR POLLICIS LONGUS

Forearm and hand—dorsal view

■ Origin
Middle third of dorsal surface of ulna, interosseous membrane

■ Insertion
Base of distal phalanx of thumb

■ Action
Extends thumb

■ Nerve
Radial nerve (C7, C8)

EXTENSORS OF THE THUMB

Forearm—posterior view

1. Abductor pollicis longus
2. Extensor pollicis brevis
3. Extensor pollicis longus

Note: Beginning at the styloid process of the radius, the tendons of the extensor pollicis longus (medially) and the extensor pollicis brevis (laterally) form a hollow depression known as the anatomical snuff box.

EXTENSOR INDICIS

Forearm and hand—dorsal view

■ Origin
Posterior surface of ulna and adjacent part of interosseous membrane

■ Insertion
Extensor expansion on dorsal surface of proximal phalanx of index finger

■ Action
Extends index finger

■ Nerve
Radial nerve (C7, C8)

PALMARIS BREVIS

Hand—palmar view

■ **Origin**
Flexor retinaculum, palmar aponeurosis

■ **Insertion**
Skin of the palm

■ **Action**
Corrugates lateral skin of palm

■ **Nerve**
Ulnar nerve (C8, T1)

ABDUCTOR POLLICIS BREVIS *(Thenar Eminence)*

Hand—palmar view

■ Origin
Tubercle of scaphoid, tubercle of trapezium, flexor retinaculum

■ Insertion
Base of proximal phalanx of thumb

■ Action
Abducts thumb and moves it anteriorly, acts together with other muscles of thenar eminence to oppose thumb to other fingers

■ Nerve
Median (C8, T1)

Note: The abductor pollicis brevis, flexor pollicis brevis, and opponens pollicis form the thenar eminence at the base of the thumb.

FLEXOR POLLICIS BREVIS *(Thenar Eminence)*

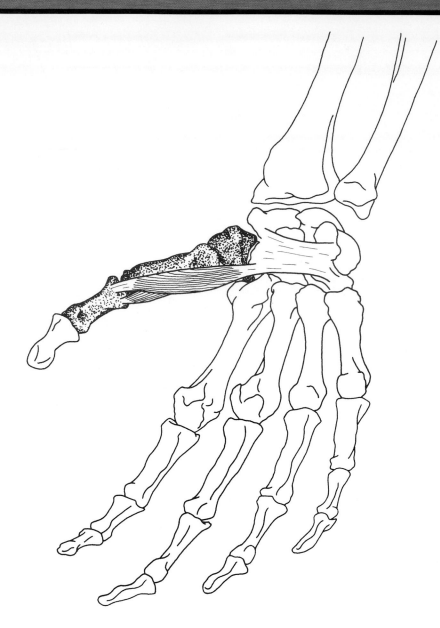

Hand—palmar view

■ Origin
Flexor retinaculum and trapezium, and first metacarpal bone

■ Insertion
Base of proximal phalanx of thumb

■ Action
Flexes metacarpophalangeal joint of thumb, assists in abduction and rotation of thumb, acts together with other muscles of thenar eminence to oppose thumb to other fingers

■ Nerve
Lateral portion—median nerve (C8, T1)

Medial portion—ulnar nerve (C8, T1)

OPPONENS POLLICIS *(Thenar Eminence)*

Hand—palmar view

■ Origin
Flexor retinaculum, tubercle of trapezium

■ Insertion
Lateral border of first metacarpal bone

■ Action
Rotates thumb into opposition with fingers, acts together with other muscles of thenar eminence to oppose thumb to other fingers

■ Nerve
Median nerve (C8, T1)

ADDUCTOR POLLICIS

Hand—palmar view

■ Origin
Oblique head—anterior surfaces of second and third metacarpals, capitate, trapezoid

Transverse head—anterior surface of third metacarpal bone

■ Insertion
Medial side of base of proximal phalanx of thumb

■ Action
Adducts thumb

■ Nerve
Ulnar nerve (C8, T1)

ABDUCTOR DIGITI MINIMI *(Hypothenar Eminence)*

Hand—palmar view

■ Origin
Pisiform bone, tendon of flexor carpi ulnaris

■ Insertion
Medial side of base of proximal phalanx of fifth finger

■ Action
Abducts fifth finger

■ Nerve
Ulnar nerve (C8, T1)

Note: The hypothenar eminence is less prominent than the thenar eminence, and the fifth finger obviously cannot oppose the other digits.

FLEXOR DIGITI MINIMI BREVIS *(Hypothenar Eminence)*

Hand—palmar view

■ **Origin**
Anterior surface of flexor retinaculum, hook of hamate

■ **Insertion**
Medial side of base of proximal phalanx of fifth finger

■ **Action**
Flexes fifth finger at metacarpophalangeal joint

■ **Nerve**
Ulnar nerve (C8, T1)

OPPONENS DIGITI MINIMI *(Hypothenar Eminence)*

Hand—palmar view

■ Origin
Anterior surface of flexor retinaculum, hook of hamate

■ Insertion
Whole length of medial border of fifth metacarpal bone

■ Action
Rotates fifth metacarpal bone, draws fifth metacarpal bone forward, assists flexor digiti minimi brevis in flexing carpometacarpal joint of fifth finger

■ Nerve
Ulnar nerve (C8, T1)

LUMBRICALES* *(Four Muscles)*

Hand—palmar view

■ Origin
Tendons of flexor digitorum profundus in palm

■ Insertion
Lateral side of corresponding tendon of extensor digitorum on fingers

■ Action
Extend fingers at interphalangeal joints, weakly flex fingers at metacarpophalangeal joints

■ Nerve
Lateral lumbricals (first and second)—median nerve (C8, T1)

Medial lumbricals (third and fourth)—ulnar nerve (C8, T1)

■ Relationship
Assist extensor digitorum communis in extending fingers without hyperextension at the metacarpophalangeal joints

*Associated with the tendons of flexor digitorum profundus.

PALMAR INTEROSSEI

Hand—palmar view

■ Origin
First—medial side of base of first metacarpal bone

Second, third, and fourth—anterior surfaces of second, fourth, and fifth metacarpal bones

■ Insertion
First—medial side of base of proximal phalanx of thumb

Second—medial side of base of proximal phalanx of index finger

Third and fourth—lateral side of proximal phalanges of ring finger and fifth finger

■ Action
Adduct fingers toward center of third finger at metacarpophalangeal joints, assist in flexion of fingers at metacarpophalangeal joints

■ Nerve
Ulnar nerve (C8, T1)

Note: The palmar interosseus of the thumb is called the palmar interosseus of Henle. Most anatomists claim that it is usually absent while some argue for its common occurrence.

DORSAL INTEROSSEI

Hand—palmar view

■ Origin
By two heads from adjacent sides of first and second, second and third, third and fourth, and fourth and fifth metacarpal bones

■ Insertion
First—lateral side of base of proximal phalanx of index finger

Second—lateral side of base of proximal phalanx of middle finger

Third—medial side of base of proximal phalanx of middle finger

Fourth—medial side of base of proximal phalanx of ring finger

■ Action
Abduct fingers away from center of third finger at metacarpophalangeal joints, assist in flexion of fingers at metacarpophalangeal joints

■ Nerve
Ulnar nerve (C8, T1)

CHAPTER EIGHT

Muscles of the Hip and Thigh

PSOAS MAJOR *(Part of Iliopsoas)*

Lumbar region, hip, and thigh—anterior view

■ Origin
Bases of transverse processes of all lumbar vertebrae, bodies of twelfth thoracic and all lumbar vertebrae, intervertebral disks above each lumbar vertebra

■ Insertion
Lesser trochanter of femur

■ Action
Flexes thigh*, flexes vertebral column

*See note on p. 188.

■ Nerve
Branches from lumbar plexus (L2, L3) and sometimes L1 or L4

Note: Some upper fibers insert onto the hip bone from the arcuate line to the iliopectineal eminence to form the *psoas minor*. This muscle has little function and is frequently absent.

ILIACUS *(Part of Iliopsoas)*

Lumbar region, hip, and thigh—anterior view

■ Origin
Upper two-thirds of iliac fossa, ala of sacrum and adjacent ligaments, anterior inferior iliac spine

■ Insertion
Onto tendon of psoas major, which continues into lesser trochanter of femur (together the two muscles form the iliopsoas)

■ Action
Flexes thigh*

■ Nerve
Femoral nerve (L2, L3)

Note: The iliacus brings swinging leg forward in walking or running.

*See note on p. 188.

PIRIFORMIS

Hip and thigh—anterior view

■ Origin
Internal surface of sacrum, sacrotuberous ligament

■ Insertion
Upper border of greater trochanter

■ Action
Laterally rotates thigh, abducts thigh

■ Nerve
Anterior rami of first and second sacral nerves

Note: The common peroneal part of the sciatic nerve may emerge through the belly of the piriformis instead of below its inferior border along with the tibial part.

OBTURATOR INTERNUS

Hip—posterior view

■ Origin
Pelvic surface of obturator membrane and surrounding bones (ilium, ischium, pubis)

■ Insertion
Common tendon with superior and inferior gemelli to medial surface of upper border of greater trochanter

■ Action
Laterally rotates thigh

■ Nerve
Nerve from sacral plexus (L5, S1)

GEMELLUS SUPERIOR

Hip—posterior view

■ Origin
Spine of ischium

■ Insertion
With tendon of obturator internus into medial surface of upper border of greater trochanter

■ Action
Laterally rotates thigh

■ Nerve
Branch of nerve to obturator internus from sacral plexus (L5, S1)

GEMELLUS INFERIOR

Hip—posterior view

■ Origin
Upper margin of ischial tuberosity

■ Insertion
With tendon of obturator internus into medial surface of upper border of greater trochanter

■ Action
Laterally rotates thigh

■ Nerve
Branch of nerve to quadratus femoris from sacral plexus

OBTURATOR EXTERNUS

Hip and thigh—anterior view

■ Origin
Outer surface of superior and inferior rami of pubis and ramus of ischium surrounding obturator foramen

■ Insertion
Trochanteric fossa of femur

■ Action
Laterally rotates thigh

■ Nerve
Obturator nerve (L3, L4)

Note: Part of this muscle can be seen posteriorly by separating the gemellus inferior and quadratus femoris. It is deep within this cleft.

QUADRATUS FEMORIS

Hip and thigh—posterior view

■ Origin
Lateral border of ischial tuberosity

■ Insertion
Below intertrochanteric crest (quadrate line)

■ Action
Laterally rotates thigh

■ Nerve
Branch from sacral plexus (L5, S1)

GLUTEUS MAXIMUS

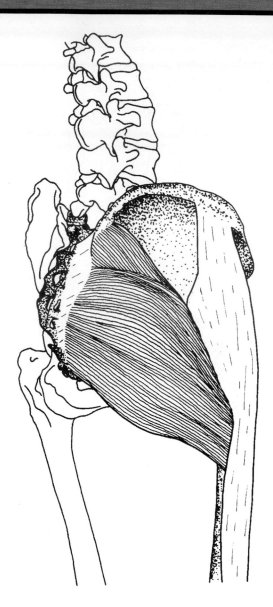

Hip and thigh—lateral view

■ Origin
Outer surface of ilium behind posterior gluteal line, adjacent posterior surface of sacrum and coccyx, sacrotuberous ligament, aponeurosis of erector spinae (sacrospinalis)

■ Insertion
Iliotibial tract of fascia lata, gluteal tuberosity of femur

■ Action
Upper part—abducts, laterally rotates thigh

Lower part—extends, laterally rotates thigh, extends trunk, assists in adduction of thigh

■ Nerve
Inferior gluteal nerve (L5, S1, S2)

Note: This is not a postural muscle; it is not used in walking but only in forceful extension, as in running, climbing, or rising from a seated position.

GLUTEUS MEDIUS

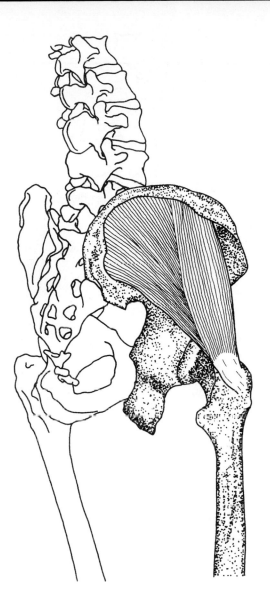

Hip and thigh—lateral view

■ Origin
Outer surface of ilium inferior to iliac crest

■ Insertion
Lateral surface of greater trochanter

■ Action
Abducts femur and rotates thigh medially

■ Nerve
Superior gluteal nerve (L4, L5, S1)

Note: In locomotion, this muscle (along with the gluteus minimus) prevents the pelvis from dropping (adduction of thigh) toward the opposite swinging leg.

GLUTEUS MINIMUS

Hip and thigh—lateral view

■ Origin
Outer surface of ilium between middle (anterior) and inferior gluteal lines

■ Insertion
Anterior surface of greater trochanter

■ Action
Abducts femur, rotates thigh medially

■ Nerve
Superior gluteal nerve (L4, L5, S1)

See note on p. 173.

MUSCLES OF THE HIP

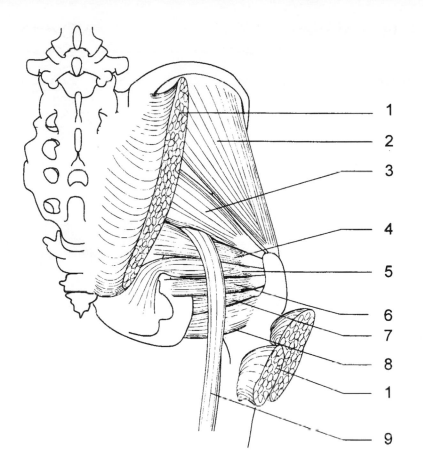

Hip—posterior view

1. Gluteus maximus (cut)
2. Gluteus medius
3. Piriformis
4. Gemellus superior
5. Obturator internus

6. Gemellus inferior
7. Obturator externus
8. Quadratus femoris
9. Sciatic nerve

Note: Gemellus inferior and quadratus femoris have been shown separated to expose the deeply placed obturator externus.

TENSOR FASCIAE LATAE

Hip and thigh—lateral view

■ Origin
Outer edge of iliac crest between anterior superior iliac spine and iliac tubercle

■ Insertion
Iliotibial tract on upper part of thigh

■ Action
Flexes, abducts thigh

■ Nerve
Superior gluteal nerve (L4, L5, S1)

Note: This muscle, along with gluteus maximus, draws the fascia lata upward, stabilizing the knee.

SARTORIUS

Hip, thigh, and leg—anterior view

■ Origin
Anterior superior iliac spine and area immediately below it

■ Insertion
Upper part of medial surface of shaft of tibia

■ Action
Flexes, abducts, and laterally rotates thigh, flexes and slightly medially rotates leg at knee joint after flexion

■ Nerve
Femoral nerve (L2, L3)

■ Relationship
Insertions of sartorius, gracilis, and semitendinosus fuse on the medial tibia; these tendons, called the pes anserinus (goose foot), give medial support to the knee

Note: This muscle is used to bring swinging leg forward in walking and running.

RECTUS FEMORIS *(One of Quadriceps Femoris)*

Hip, thigh, and leg—anterior view

■ Origin

Anterior head—anterior inferior iliac spine

Posterior head—ilium above acetabulum

■ Insertion

Patella, then by patellar ligament to tuberosity of the tibia

■ Action

Extends leg at knee joint, flexes thigh

■ Nerve

Femoral nerve (L2–L4)

Note: This muscle is used when thigh flexion and leg extension are needed together, such as in kicking a football. In walking, the quadriceps prevent the knee from flexing during heel strike and early support phase.

VASTUS LATERALIS *(One of Quadriceps Femoris)*

Hip, thigh, and leg—anterior view

■ Origin
Intertrochanteric line, inferior border of greater trochanter, gluteal tuberosity, lateral lip of linea aspera of femur

■ Insertion
Lateral margin of patella, then by patellar ligament to tuberosity of tibia

■ Action
Extends leg

■ Nerve
Femoral nerve (L2–L4)

VASTUS MEDIALIS *(One of Quadriceps Femoris)*

Hip, thigh, and leg—anterior view

■ Origin
Intertrochanteric line, medial lip of linea aspera of femur, medial intermuscular septum of adductor magnus and adductor longus, medial supracondylar ridge

■ Insertion
Medial border of the patella, then by patellar ligament into tibial tuberosity, medial condyle of tibia

■ Action
Extends leg, draws patella medially

■ Nerve
Femoral nerve (L2–L4)

Note: Although they are not anatomically separate[1], the lower portion of the vastus medialis is referred to as the vastus medialis obiliquus (VMO). Its action is to stabilize the patella and prevent its lateral dislocation.

[1]Noric, M. M., Mitchell, J., de Klerk, D. (1997) A comparison of the proximal and distal parts of the vastus medialis muscle. *Australian Journal of Physiotherapy,* 43(4):277–281.

VASTUS INTERMEDIUS *(One of Quadriceps Femoris)*

Hip, thigh, and leg—anterior view

■ Origin
Anterior and lateral surfaces of upper two-thirds of femur, lateral intermuscular septum, linea aspera, lateral supracondylar ridge

■ Insertion
Deep aspect of quadriceps tendon, then through patella to tibial tuberosity

■ Action
Extends leg

■ Nerve
Femoral nerve (L2–L4)

Note: A few bundles of fibers from this muscle insert onto the upper part of the joint capsule of the knee. They probably draw the capsule superiorly during extension of the leg, preventing it from binding in the joint. They are called *articularis genus*.

MUSCLES OF THE ANTERIOR THIGH

Hip and thigh—anterior view

1. Tensor fasciae latae
2. Iliotibial tract
3. Vastus lateralis (quadriceps femoris)
4. Vastus intermedius (quadriceps femoris)
5. Rectus femoris (cut) (quadriceps femoris)
6. Sartorius
7. Vastus medialis (quadriceps femoris)

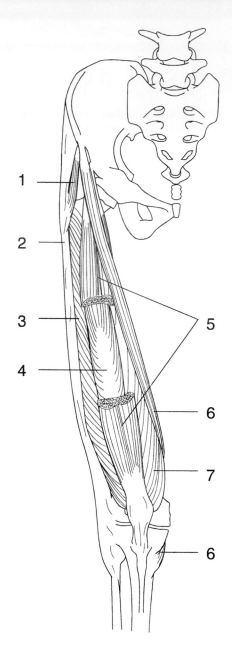

BICEPS FEMORIS *(Part of Hamstrings)*

Hip and thigh—posterior view

■ Origin
Long head—ischial tuberosity, sacrotuberous ligament

Short head—linea aspera, lateral supracondylar ridge, lateral intermuscular septum

■ Insertion
Lateral side of head of fibula and lateral condyle of tibia

■ Action
Flexes leg, long head also extends thigh

■ Nerve
Long head—tibial part of sciatic nerve (S1–S3)

Short head—common peroneal part of sciatic nerve (L5, S1, S2)

Note: During walking or running, the hamstrings are used to slow down the leg at the end of its swing and prevent the trunk from flexing at the hip. They are susceptible to being strained by resisting the momentum of these body parts.

SEMITENDINOSUS *(Part of Hamstrings)*

Hip and thigh—posterior view

■ Origin
Ischial tuberosity

■ Insertion
Medial surface of shaft of tibia

■ Action
Flexes and slightly medially rotates leg; after flexion, extends thigh

■ Nerve
Tibial portion of sciatic nerve (L5, S1, S2)

See note on p. 183 and Relationship section on p. 177.

SEMIMEMBRANOSUS *(Part of Hamstrings)*

Hip and thigh—posterior view

■ **Origin**
Ischial tuberosity

■ **Insertion**
Posterior part of medial condyle of tibia

■ **Action**
Flexes and slightly medially rotates leg; after flexion, extends thigh

■ **Nerve**
Tibial portion of sciatic nerve (L5, S1, S2)

See note on p. 183.

HAMSTRING MUSCLES

Hip and thigh—posterior view

1. Sciatic nerve
2. Quadratus femoris
3. Biceps femoris
4. Semimembranosus
5. Semitendinosus
6. Tibial nerve
7. Common peroneal nerve

Note: The common peroneal nerve is exposed to compression and damage as it passes over the head of the fibula. The quadratus femoris, a lateral rotator, is included for reference.

GRACILIS

Hip and thigh—anterior view

■ Origin
Lower margin of body and inferior ramus of pubis

■ Insertion
Upper part of medial surface of shaft of tibia

■ Action
Adducts thigh at hip joint and flexes leg, with leg flexed, assists in medial rotation

■ Nerve
Obturator nerve (L2–L4)

See Relationship section on p. 177.

PECTINEUS

Hip and thigh—anterior view

■ Origin
Pectineal line on superior ramus of pubis

■ Insertion
From lesser trochanter to linea aspera of femur

■ Action
Flexes thigh, assists in adduction when hip is flexed

■ Nerve
Femoral nerve (L2, L3), (sometimes a branch of obturator nerve)

Note: The rotational component of thigh muscle action depends upon the starting position of the hip joint. The pectineus, adductor longus, adductor brevis, and psoas major probably assist in medial rotation when the thigh is in anatomical position but may shift to assisting in lateral rotation as the thigh flexes and abducts.

The iliacus and adductor tubercle part of adductor magnus probably assist in medial rotation throughout the range of motion of the hip joint while the linea aspera part of the adductor magnus may be a slight lateral rotator.

ADDUCTOR LONGUS

Hip and thigh—posterior view

■ **Origin**
Anterior of body of pubis

■ **Insertion**
Medial lip of linea aspera

■ **Action**
Adducts, flexes thigh, assists in medial rotation*

■ **Nerve**
Obturator nerve (L2–L4)

*See note on p. 188.

ADDUCTOR BREVIS

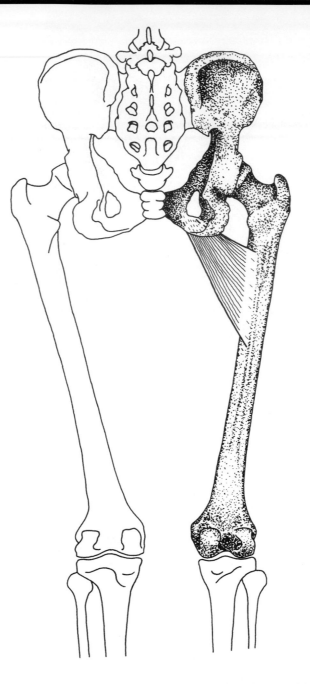

Hip and thigh—posterior view

■ **Origin**

Outer surface of inferior ramus of pubis

■ **Insertion**

From below lesser trochanter to linea aspera and into proximal part of linea aspera

■ **Action**

Adducts thigh, assists in flexion, medial rotation*

■ **Nerve**

Obturator nerve (L2–L4)

*See note on p. 188.

ADDUCTOR MAGNUS

Hip and thigh—posterior view

■ Origin
Inferior ramus of pubis, ramus and lower part of tuberosity of ischium

■ Insertion
Linea aspera, adductor tubercle of femur

■ Action
Adducts, extends thigh, lower portion (adductor tubercle insertion) assists in medial rotation*

■ Nerve
Obturator nerve (L2–L4), sciatic nerve

*See note on p. 188.

HIP FLEXORS AND ADDUCTORS

Hip and thigh—anterior view

1. Psoas major
2. Iliacus
3. Inguinal ligament
4. Femoral nerve, vein, artery
5. Pectineus
6. Adductor brevis
7. Adductor longus (cut)
8. Adductor magnus
9. Gracilis

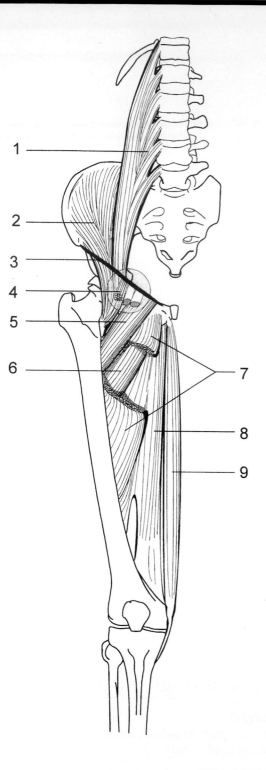

CHAPTER NINE

Muscles of the Leg and Foot

TIBIALIS ANTERIOR

Leg—anterolateral view

■ Origin
Lateral condyle of tibia, upper half of lateral surface of tibia, interosseous membrane

■ Insertion
Medial side and plantar surface of medial cuneiform bone, base of first metatarsal bone

■ Action
Dorsiflexes foot, inverts (supinates) foot

■ Nerve
Deep peroneal nerve (L4, L5, S1)

EXTENSOR HALLUCIS LONGUS

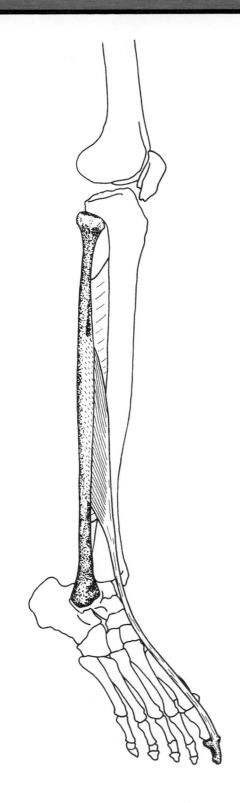

Leg—anterolateral view

■ Origin
Middle half of anterior surface of fibula and interosseous membrane

■ Insertion
Base of distal phalanx of great toe

■ Action
Extends, hyperextends great toe, dorsiflexes and inverts (supinates) foot

■ Nerve
Deep peroneal nerve (L4, L5, S1)

EXTENSOR DIGITORUM LONGUS

Foot—anterolateral view

■ Origin
Upper two-thirds of anterior surface of fibula, interosseous membrane, lateral condyle of tibia

■ Insertion
Along dorsal surface of four lateral toes, and then to bases of middle and distal phalanges

■ Action
Extends toes, dorsiflexes foot, everts foot

■ Nerve
Deep peroneal nerve (L4, L5, S1)

Note: The lower lateral part of this muscle makes a separate insertion onto the dorsal surface of the fifth metatarsal and is called *peroneus tertius.*

Leg—anterolateral view

FIBULARIS TERTIUS *(Peroneus Tertius)*

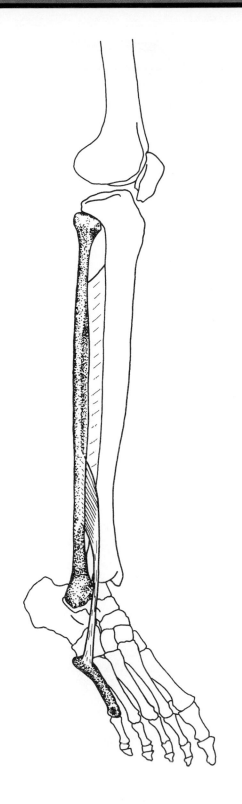

Leg—anterolateral view

■ Origin
Lower third of anterior surface of fibula and interosseous membrane

■ Insertion
Dorsal surface of base of fifth metatarsal bone

■ Action
Dorsiflexes and everts foot

■ Nerve
Deep peroneal nerve (L4, L5, S1)

Note: This muscle is not present in all individuals. It may be described as the fifth tendon of extensor digitorum longus.

ANTERIOR AND LATERAL LEG MUSCLES

Leg—anterolateral view

1. Fibularis longus
2. Fibularis brevis
3. Fibularis tertius
4. Tibialis anterior
5. Extensor retinaculum
6. Extensor hallucis longus
7. Extensor digitorum longus
8. Extensor digitorum brevis

GASTROCNEMIUS *(Part of Triceps Surae)*

Leg—posterior view

■ Origin
Lateral head—lateral condyle and posterior surface of femur

Medial head—popliteal surface of femur above medial condyle

■ Insertion
Posterior surface of the calcaneus

■ Action
Plantar flexes foot, flexes leg when foot is dorsiflexed and knee is extended

■ Nerve
Tibial nerve (S1, S2)

SOLEUS *(Part of Triceps Surae)*

Leg—posterior view

■ Origin
Posterior surface of the tibia (soleal line), upper third of posterior surface of fibula, fibrous arch between tibia and fibula

■ Insertion
Posterior surface of the calcaneus

■ Action
Plantar flexes foot

■ Nerve
Tibial nerve (S1, S2)

PLANTARIS

Leg—posterior view

■ Origin
Lateral supracondylar ridge of femur, oblique popliteal ligament

■ Insertion
Posterior surface of the calcaneus

■ Action
Plantar flexes foot, flexes leg

■ Nerve
Tibial nerve (S1, S2)

MUSCLES OF THE CALF

Leg—posterior view

1. Soleus
2. Plantaris
3. Gastrocnemius (cut)

POPLITEUS

Leg—posterior view

■ Origin
Lateral surface of lateral condyle of femur

■ Insertion
Upper part of posterior surface of tibia

■ Action
Rotates leg medially, flexes leg

■ Nerve
Tibial nerve (L4, L5, S1)

Note: Stern contends that this muscle stabilizes the knee by preventing lateral rotation of the tibia during medial rotation of the thigh while the foot is planted. Knee stabilization allows standing without undue fatigue to the quadriceps.

Reference: Stern, J. T. *Essentials of Gross Anatomy*, F. A. Davis Company, Philadelphia, 1988.

FLEXOR HALLUCIS LONGUS

Leg—posterior view

■ Origin
Lower two-thirds of posterior surface of shaft of fibula, posterior intermuscular septum, interosseous membrane

■ Insertion
Base of distal phalanx of great toe

■ Action
Flexes distal phalanx of great toe, assists in plantar flexing foot, inverts foot

■ Nerve
Tibial nerve (L5, S1, S2)

Note: This muscle is important in pushing off the surface in walking, running, jumping.

FLEXOR DIGITORUM LONGUS

Foot—plantar view

■ **Origin**
Medial part of posterior surface of tibia

■ **Insertion**
Bases of distal phalanges of second, third, fourth, and fifth toes

■ **Action**
Flexes distal phalanges of lateral four toes, assists in plantar flexing foot, inverts foot

■ **Nerve**
Tibial nerve (L5, S1)

Leg—posterior view

TIBIALIS POSTERIOR

Foot—plantar view

■ Origin
Lateral part of posterior surface of tibia, interosseous membrane, proximal half of posterior surface of fibula

■ Insertion
Tuberosity of navicular bone, cuboid, cuneiforms, second, third, and fourth metatarsals, sustentaculum tali of calcaneus

■ Action
Plantar flexes, inverts foot

■ Nerve
Tibial nerve (L5, S1)

Leg—posterior view

DEEP POSTERIOR LEG MUSCLES

Leg—posterior view

1. Popliteus
2. Tibialis posterior
2a. Tendon of tibialis posterior
3. Flexor hallucis longus (cut)
4. Flexor digitorum longus (cut)
5. Medial malleolus

Note: The order of the three tendons, from anterior to posterior, passing posterior to the medial malleolus can be remembered by the following mnemonic: Tom (*tibialis* posterior), Dick (flexor *digitorum*), Harry (flexor *hallucis* longus).

FIBULARIS LONGUS *(Peroneus Longus)*

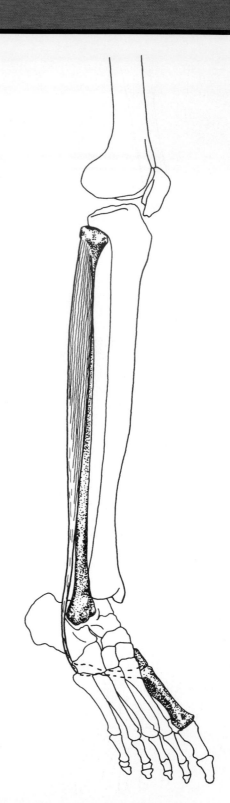

Foot—plantar view

■ **Origin**
Upper two-thirds of lateral surface of fibula

■ **Insertion**
Lateral side of medial cuneiform, base of first metatarsal

■ **Action**
Plantar flexes, everts foot

■ **Nerve**
Superficial peroneal nerve (L4, L5, S1)

Leg—anterolateral view

FIBULARIS BREVIS *(Peroneus Brevis)*

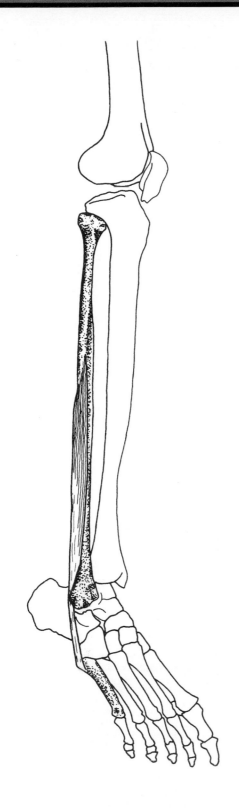

Leg—anterolateral view

■ Origin
Lower two-thirds of lateral surface of fibula

■ Insertion
Lateral side of base of fifth metatarsal bone

■ Action
Everts, plantar flexes foot

■ Nerve
Superficial peroneal nerve (L5, S1, S2)

EXTENSOR DIGITORUM BREVIS

Foot—anterolateral view

■ Origin
Anterior and lateral surfaces of calcaneus, lateral talocalcaneal ligament, inferior extensor retinaculum

■ Insertion
Into base of proximal phalanx of great toe; into lateral sides of tendons of extensor digitorum longus of second, third, and fourth toes

■ Action
Extends the four toes

■ Nerve
Deep peroneal nerve (S1, S2)

ABDUCTOR HALLUCIS *(First Layer)*

Foot—plantar view

■ Origin
Tuberosity of calcaneus, flexor retinaculum, plantar aponeurosis

■ Insertion
Medial side of base of proximal phalanx of great toe

■ Action
Stabilizes great toe (with adductor hallucis)

■ Nerve
Medial plantar nerve (L4, L5)

Note: The muscles of the sole of the foot can be divided into four layers (from superficial to deep):

First layer—abductor hallucis, flexor digitorum brevis, abductor digiti minimi

Second layer—quadratus plantae, lumbricales (tendons of flexor hallucis longus and flexor digitorum longus pass through this layer)

Third layer—flexor hallucis brevis, adductor hallucis, flexor digiti minimi brevis

Fourth layer—interossei (tendons of tibialis posterior and peroneus longus pass through this layer)

FLEXOR DIGITORUM BREVIS *(First Layer)*

Foot—plantar view

■ Origin
Tuberosity of calcaneus, plantar aponeurosis

■ Insertion
Sides of middle phalanges of second to fifth toes

■ Action
Flexes proximal phalanges and extends distal phalanges of second through fifth toes

■ Nerve
Medial plantar nerve (L4, L5)

ABDUCTOR DIGITI MINIMI *(First Layer)*

Foot—plantar view

■ **Origin**
Tuberosity of calcaneus, plantar aponeurosis

■ **Insertion**
Lateral side of proximal phalanx of fifth toe

■ **Action**
Abducts fifth toe

■ **Nerve**
Lateral plantar nerve (S1, S2)

QUADRATUS PLANTAE *(Second Layer)*

Foot—plantar view

■ **Origin**

Medial head—medial surface of calcaneus

Lateral head—lateral border of inferior surface of calcaneus

■ **Insertion**

Lateral margin of tendon of flexor digitorum longus

■ **Action**

Flexes terminal phalanges of second through fifth toes

■ **Nerve**

Lateral plantar nerve (S1, S2)

LUMBRICALS *(Second Layer)*

Foot—plantar view

■ **Origin**
Tendons of flexor digitorum longus

■ **Insertion**
Dorsal surfaces of proximal phalanges

■ **Action**
Flex proximal phalanges of second through fifth toes

■ **Nerve**
First lumbricalis—medial plantar nerve (L4, L5)

Second through fifth lumbricales—lateral plantar nerve (S1, S2)

FLEXOR HALLUCIS BREVIS *(Third Layer)*

Foot—plantar view

■ Origin
Cuboid bone, lateral cuneiform bone

■ Insertion
Medial part—medial side of base of proximal phalanx of great toe

Lateral part—lateral side of base of proximal phalanx of great toe

■ Action
Flexes proximal phalanx of great toe

■ Nerve
Medial plantar nerve (L4, L5, S1)

Note: The tendons of insertion contain sesamoid bones.

ADDUCTOR HALLUCIS *(Third Layer)*

Foot—plantar view

■ Origin
Oblique head—second, third, and fourth metatarsal bones, and sheath of peroneus longus tendon

Transverse head—plantar metatarsophalangeal ligaments of third, fourth, and fifth toes, and transverse metatarsal ligaments

■ Insertion
Lateral side of base of proximal phalanx of great toe

■ Action
Stabilizes great toe (with abductor hallucis)

■ Nerve
Lateral plantar nerve (S1, S2)

FLEXOR DIGITI MINIMI BREVIS *(Third Layer)*

Foot—plantar view

■ Origin
Base of fifth metatarsal, sheath of peroneus longus tendon

■ Insertion
Lateral side of base of proximal phalanx of fifth toe

■ Action
Flexes proximal phalanx of fifth toe

■ Nerve
Lateral plantar nerve (S1, S2)

DORSAL INTEROSSEI *(Fourth Layer; Four Muscles)*

Foot—anterolateral view

■ Origin
Adjacent sides of metatarsal bones

■ Insertion
Bases of proximal phalanges

First—medial side of proximal phalanx of second toe

Second, third, fourth—lateral sides of proximal phalanges of second, third, and fourth toes

■ Action
Abduct toes, flex proximal phalanges

■ Nerve
Lateral plantar nerve (S1, S2)

PLANTAR INTEROSSEI *(Fourth Layer; Three Muscles)*

Foot—plantar view

■ Origin
Bases and medial sides of third, fourth, and fifth metatarsal bones

■ Insertion
Medial sides of bases of proximal phalanges of same toes

■ Action
Adduct toes, flex proximal phalanges

■ Nerve
Lateral plantar nerve (S1, S2)

Alphabetical Listing of Muscles

Index

Notes

Notes

Notes

Notes

Notes